CANALS *and American Economic Development*

CANALS *and American* Economic *Development*

BY

CARTER GOODRICH

JULIUS RUBIN

H. JEROME CRANMER

HARVEY H. SEGAL

Edited by
CARTER GOODRICH

IRA J. FRIEDMAN DIVISION
KENNIKAT PRESS
Port Washington, N. Y./London

ERRATA

Page 205, line 28, *for* multiple, *read* multiplier.
Page 243, line 29, *for* $3.1, *read* $12.5; line 30, *for* two thirds, *read* $2 million; line 33, *for* .36, *read* 3.6.
Page 244, line 21, *for* Table 8, *read* Table 12.
Page 290, line 34, *for* $2.0 + 1.05 = $3.05 million, *read* $2.0 + 10.5 = $12.5 million.

CANALS AND AMERICAN ECONOMIC DEVELOPMENT

Copyright © 1961 by Columbia University Press
Reissued in 1972 by Kennikat Press by arrangement
Library of Congress Catalog Card No.: 72-86271
ISBN 0-8046-1765-1

Manufactured by Taylor Publishing Company   Dallas, Texas

Foreword

This study of early American development policy is the first publication of Columbia University's Graduate Workshop on the Economic Development of the Industrial Countries. Its subject is appropriate to the general theme of the Workshop's work, which is a reexamination of the economic history of the developed industrial areas of the world, particularly Western Europe and the United States, in the light of modern discussions of the problem of economic development.

Publication was made possible by the generous gift of the Ford Foundation for the establishment of a group of Graduate Economics Workshops, intended to improve the conditions of work for graduate students of economics at Columbia by providing opportunities for research and discussion and for contact with older scholars engaged on related tasks.

The original plan of the work was simply to make available in a single convenient volume the principal findings of three Columbia dissertations—one by Jerome Cranmer on The New Jersey Canals (1955), the second by Harvey Segal on Canal Cycles (1956) and the third by Julius Rubin on Imitation by Canal or Innovation by Railroad (1959)—together with the quantitative evidence on canal investment presented by Mr. Cranmer and discussed by Mr. Segal at the 1957 meeting of the Conference on Income and Wealth. Production of a unified book, however, appeared to require somewhat more than this; and new research was undertaken so that Chapter I could present the early policies of New York State for comparison with those of Pennsylvania and New Jersey and so that Chapter V could attempt to assess the impact of the canal movement on the economic development of the United States.

The volume is the product of joint planning and close collaboration throughout, but Mr. Rubin is primarily responsible for

Chapters I and II, Mr. Cranmer for Chapter III and the section "Estimates of Canal Construction, 1815–60" at the end of Chapter IV, and Mr. Segal for Chapters IV and V. The editor wrote the Introduction and Conclusion, and his book on *Government Promotion of American Canals and Railroads*, published in 1960 by Columbia University Press, provided part of the background for the study.

Sections of the manuscript were submitted, in earlier form, for discussion at meetings of the Industrial Countries Workshop, and the writers profited by the vigorous and searching criticisms made by its members. Suggestions by Professor Joseph Dorfman and Mr. David Novack, of the Columbia University Department of Economics, proved particularly helpful. The Secretary of the Workshops, Miss Elinor Ricker, has kept a firm and friendly hand on the management of the operation. As in the case of *Government Promotion*, Mrs. Kathryn W. Sewny served with skill and understanding as editor on behalf of the Press.

Research for Mr. Segal's dissertation received support from the Committee (now Council) on Research in Economic History, as did that of Mr. Cranmer for a briefer period; and Mr. Rubin held a Dissertation Fellowship awarded by the Ford Foundation. For advice on his contribution to the present volume, Mr. Segal wishes to thank Professors Gerhard Bry, Herman E. Krooss, and Clifford D. Clark of New York University, and Mr. Arnold Faden.

Acknowledgement should also be made of the courtesy of the National Bureau of Economic Research, the Conference on Research in Income and Wealth, and the Princeton University Press in allowing us to use the substance of the Cranmer and Segal contributions to the 1957 Conference, published in *Trends in the American Economy in the Nineteenth Century (Studies in Income and Wealth* XXIV; Princeton, 1960), and of the American Philosophical Society in permitting Mr. Rubin to use the findings of Part I of his volume, *Canal or Railroad,* in the Society's *Transcriptions* (Philadelphia, 1961).

CARTER GOODRICH

Columbia University
May 10, 1961

Contents

Introduction, BY CARTER GOODRICH   1

THE POLITICAL DECISIONS

I. An Innovating Public Improvement: The Erie Canal, BY JULIUS RUBIN   15

II. An Imitative Public Improvement: The Pennsylvania Mainline, BY JULIUS RUBIN   67

III. Improvements Without Public Funds: The New Jersey Canals, BY H. JEROME CRANMER   115

THE ECONOMIC IMPACT

IV. Cycles of Canal Construction, BY HARVEY H. SEGAL   169
    Estimates of Canal Investment, 1815–60   208
    Public Funds in Canal Construction, 1815–60   213

V. Canals and Economic Development, BY HARVEY H. SEGAL   216
    The Demand for Canal Transport Services and Benefits Conferred   247

Conclusion, BY CARTER GOODRICH   249

Notes   256

Index   293

Map

Principal Canals of the Ante-Bellum Period   184

Figures

1 Canal Cycles, 1815–60   173
2 The Demand for Canal Transport Services and Benefits Conferred   248

Tables

1 Cycles of Canal Construction, 1815–60   172
2 Canal Investment in the United States, 1815–60   208
3 Observed and Estimated Investment by Canal Cycles, 1815–60   210
4 Canals Included in the Annual Estimates of Investments, 1815–60   211
5 Estimates of Public Aid to Canals, 1815–60   214
6 Public Investment in Canals, 1815–60   215
7 Growth of National Commodity Output at Constant Prices and Traffic from the West on the Erie Canal, 1839–59   229
8 Distribution of the Commodity Trade of the Northwest by Shipping Routes, 1835–53   231
9 Canal Investment and State Income Generated in Commodity Production and Distribution, 1840   233
10 Changes in Occupational Distribution of Employment in Erie Canal Counties and Others, New York, 1820–40   236
11 Changes in Occupational Distribution of Employment in Canal and Non-Canal Counties, Ohio, 1820–40   237
12 Average Ton-Mileage on Ten Heavily Utilized Canals, 1837–46   242
13 Investment in Canals That Failed   246

Introduction

BY CARTER GOODRICH

This is a set of connected studies of the planning of economic development. The extensive programs by which American governments in the nineteenth century promoted internal improvements were clearly developmental in purpose, though "economic development" had not yet become a slogan. Before the coming of the railroad the most important of these improvements were those which substituted the canal boat for the wagon or pack train, particularly in the trade between East and West. For the construction of the nation's canals initiative and capital came mainly from governmental sources, and their triumphs and failures provide an early and significant case of economic planning in American history.

As a study of planning, the book is concerned with the decisions to construct the canals. As a study of development, it attempts to analyze their influence on the growth of the American economy. Our interest, like that of Robert Fulton in 1796, is in "canal navigation as connected with political economy." [1]

When Americans in Fulton's day began to plan and build canals, the principal model for a major project was still the French Canal of Languedoc, which connected the Atlantic with the Mediterranean by a cut of nearly 150 miles requiring more than a hundred locks. Constructed under the orders of Louis XIV, it remained, to Fulton's republican eyes, "the noblest monument of the monarch who patronized it." [2] Yet Americans paid much more attention to the great burst of activity in canal building in which Great Britain was then engaged. "Within the short space of half a century," as an admiring French observer reported in 1825, "a double row of canals is formed . . . uniting together opposite seas; basins separated by numberless chains of hills and mountains; opulent ports; industrious towns; fertile plains; and inexhaustible mines." With more than two thousand miles of canals and extensive stretches of improved or canalized river navigation, British construction had far surpassed the French figure.

Observers were also impressed by the manner in which this rapid progress had been made and by the financial results. Baron Dupin asked rhetorically what the British government had done to produce these great public monuments and answered with the one word: "Nothing." "Nothing," he added, except to give free rein to commerce and provide "protection without, liberty within, and justice everywhere." [3] Though the answer overlooked loans made by the Exchequer to several of the English companies and more substantial government effort in Scotland, most of the canals were in fact built entirely by private means. The innovating improvement was the canal constructed by that "spirited and patriotic nobleman," the Duke of Bridgewater, to carry coal from his mines to Manchester.[4] His experiment was a great success, giving him, said Fulton, "immortality and $130,000 a year." [5] Almost all the later canals were built by joint stock companies. Some eighteen of the companies paid dividends in 1825 ranging from 6 percent to 75 percent "plus bonus." [6] Though there were also a number of financial failures, profit figures like these were a powerful stimulus to the development of what was often called a "canal mania" in Great Britain, and they were cited again and again by those who were anxious to stimulate the Americans to similar achievement.

This progress had been achieved only by the conquest of substantial physical difficulties, greater, it has been suggested, than those which some decades later were to confront the builders of railroads.[7] The basic innovation—the canal lock—was known to the Dutch and the Italians even before it was used by Leonardo da Vinci. There remained, however, many problems as to where the locks should be placed and how they should be built and operated. The alternative of the inclined plane was occasionally employed. Assurance of an adequate supply of water at all seasons, particularly at the summit level, called for careful judgment as to location and often required the construction of an elaborate system of feeders and reservoirs. The digging of the ditch itself was for the most part done by hand labor with simple tools, but the difficulty varied with differences in the nature of the ground. "Puddling" the bottom to prevent the escape of the water was often a somewhat complicated affair. Still more difficult was the problem of protecting the sides of the canal against erosion by rain or floods, against the depredations of muskrats, and against the wash from canal boats.

These and other difficulties were for the most part met by practical men who developed their techniques on the job itself. "Nature alone made a Geometer" of M. Riquet, who constructed the Canal of Languedoc;[8] and Joseph Brindley, who built the Duke's canal and a number of others, was entirely self-taught. By the early decades of the nineteenth century, however, a number of treatises and pamplets on canal building had appeared, and England and France each possessed a small corps of professionally trained engineers. At the time when Americans first began to take a serious interest in the building of canals, two issues were under particularly active discussion. The first was one of size: Should canals be built with small dimensions at relatively low cost, "meandering the fields," as Fulton pictured them, and "bending their branches round each hill";[9] or should they be built on a more substantial scale to permit the economies of carriage in larger boats? The second concerned the alternatives of continuous canals as against river improvements: Should the natural bed of the river be used wherever possible, with canals resorted to only where falls or rapids must be passed; or should full canals be constructed instead? On this question, river improvements had often appeared to be the obvious solution, but the weight of professional opinion favored the greater dependability of the continuous canal.

In the United States, the early canal builders confronted still greater technical and financial difficulties. Neither the terrain nor the economy of the new country reproduced English conditions. By far the greater part of the British canal system, as Dupin pointed out, was confined to a square of less than 180 miles to the side, of which the four corners were represented by Liverpool, Hull, London, and Bristol.[10] Within this limited area, population was dense, production was concentrated, and markets and lines of trade were already well established. The "chains of hills and mountains" imposed no insuperable barriers. Though canals crossed the central Pennine range at three points, the highest altitude reached was 650 feet.[11] By contrast, America was a land of greater distances, higher mountains, severer climate, and much sparser population. To connect the Hudson River with Lake Erie would require a canal twice the length of Dupin's British square, and other projects contemplated crossing the Appalachian Mountains at altitudes of more than 2,000 feet. The terrain offered a greater variety of obstacles, and the builders of one great canal boasted of their triumph over

> the rude and undulating surface which was traversed, the rocks which were to be blasted, the irregular ledges filled with chasms and fissures which were to form the sides and basis of a water-tight canal; the spungy swamps, and gravel beds, and quicksands, which were to be made impervious to water.[12]

No country of western Europe had had to face the problem of organizing and maintaining construction forces in wilderness areas. "In America where Manual Labor" was, as Fulton pointed out, so scarce and "of so much importance," it was particularly urgent to develop mechanical methods for clearing the land and for excavation.[13] In the new country, moreover, there were virtually no professional engineers and no facilities for training them until the United States Military Academy began to supply the deficiency in the 1820s.[14]

Under American conditions, few projectors of canals were able to follow the British precedent of the profitable private corporation. Only a very small number of projects offered prospects of quick returns based on trade between established centers. The great majority were developmental in character, reaching out into areas still largely unsettled. Substantial traffic could not be expected until the improvements themselves stimulated development, and much of the ultimate gain would not be collectible in the tolls of the canals. The new country lacked "spirited noblemen" and—more important—the abundance of funds for investment that filled the subscription lists of the British companies. Domestic capital was scarce and could command a high return; and no American corporations, except for a handful of banks and insurance companies, could be confident of their ability to obtain capital from abroad. Complete or even major reliance on private enterprise therefore appeared impossible, and appeals for public support were reinforced by considerations of defense and national unity and sometimes by the conviction that enterprises of such great importance to the people as a whole should not be left in the hands of private monopolists.

Decision that American canal building must be done largely with public funds did not of itself determine which governments should be used or what methods they should employ. It was the national government that produced the most comprehensive plan. The *Report* on roads and canals, presented to the Senate in 1808 by Albert Gallatin, Secretary of the Treasury, proposed the creation of an intercoastal waterway by cutting short canals across four "necks of land" along the Atlantic coast and a series of measures for connecting the seaboard with the interior. Advantage should be taken of the great opening made by the Hudson and Mohawk rivers by building canals to Lake Champlain and Lake Ontario, with a third to by-pass Niagara Falls. Further south, each of four great river systems—the Susquehanna, the Potomac, the James, and either the Santee or the Savannah—was to be improved by canal and otherwise, and a portage road built from the head of navigation to the nearest river on the western side. Though the federal government might either do the work on its own account or in partnership with private enterprise, it must—in Gallatin's view—take the responsibility for the selection of routes so that they would conform to the great geographic features of the country rather than to the dictates of local interest.[15] Unfortunately for the success of his proposals, the surplus revenues, on which he counted as the source of funds, disappeared as the War of 1812 approached; and the federal government was never to take the major role in ante-bellum canal construction. It built the short Carondelet Canal near New Orleans as a war measure; it subscribed a million dollars to the stock of the Chesapeake and Ohio Canal Company and smaller amounts to three other enterprises; and it aided canal building in the Old Northwest by grants of land from the public domain. But all attempts to carry out a more comprehensive national program foundered on the rocks of state and sectional rivalries.[16]

A much greater role was played by the government of the states. The public works programs of New York, Pennsylvania, and Ohio alone were responsible for more than half of the entire canal investment before the Civil War.[17] To these should be added the substantial sums spent by Indiana on the Wabash and Erie, by Illinois on the Illinois and Michigan, by Maryland on the Chesapeake and Ohio after it had become virtually a state enterprise, and by Virginia on the James River and Kanawha during its period of state ownership. In still other cases, state governments contributed to canal construction by subscriptions to mixed enterprise or by loans to private companies. So also did a number of ambitious cities, though municipal investment in canals did not approach the amounts that were currently subscribed by Baltimore and Philadelphia to competing railroads.

The private share of total canal investment during this period was somewhat more than a quarter of the total. The early Santee and Middlesex canals, the coal-carrying Lehigh Canal in Pennsylvania, and two New Jersey enterprises which we shall examine in detail were built entirely without public funds. But there were very few others, and by far the greater part of private investment was made in partnership with federal, state, or local authorities.

The major beginning of actual canal building was the construction of the Erie Canal as an enterprise of the state of New York. Begun in 1817 and completed in 1825, joining the Hudson

Introduction

River with Lake Erie by a single canal, this effected the connection between seaboard and interior by a bolder stroke than Gallatin contemplated. Contemporary discussion leaves no doubt that this was the decisive turning point. Before the Erie Canal was begun, said George Armroyd in 1826, "there was not, perhaps, 100 miles of canal-work finished in the United States." The work completed included the Middlesex and Santee canals, by which the merchants of Boston reached out to capture the trade of the Merrimack Valley and their colleagues in Charleston, South Carolina, reached out to the trade of the Santee river system. The remaining mileage "was composed chiefly of very short canals, at some of the rapids and falls of rivers." Yet within a year of the completion of the Erie, Armroyd was able to list a hundred projects "either existing, or in progress of construction, or in contemplation" which would have given the United States more than 6,500 miles of canal. Not all of this contemplation was translated into action, but events came close to justifying the author's prediction that some 3,000 miles would be in operation within ten years.[18] His figure was reached in 1840, and less than a quarter century of active American building had produced a mileage considerably greater than that attained during the great half century of British construction. Later projects added another 1,000 miles. Abandonments of canal mileage began in the 1840s, and during the 1850s they exceeded the amount of new construction, though considerable sums continued to be invested both in new ventures and in the enlargement of existing canals. Abandonment was mainly the result of railroad competition, but the major era of canal building did not come to an end until most of the apparent opportunities offered by the American terrain had already been exploited.[19]

These opportunities were of various sorts. Three major efforts were made to breach the Appalachian Barrier and to compete with the Erie Canal in providing access to the West. Each of them followed one of the river routes indicated in the Gallatin report. The Pennsylvania State Works made use of the Susquehanna and Juniata valleys on the east and that of the Allegheny River on the west. The Chesapeake and Ohio Canal was designed to exploit the Potomac route, and the objective of the James River and Kanawha Canal Company was indicated by its title. None of the three fulfilled its original ambitious design of carrying a canal over—or through—the Alleghenies. Pennsylvania accomplished the mountain crossing by means of the Portage Railroad, and its combined rail-and-water system was for the most part superseded by the all-rail Pennsylvania Railroad in the fifties. The Chesapeake and Ohio Canal established its terminus at Cumberland, at the foot of the Alleghenies, while the Baltimore and Ohio Railroad went on to cross the mountains. The James River and Kanawha still hoped on the eve of the Civil War—and even after—to achieve a full canal to the West and continued to seek support in competition with the rival project of a railroad.

Another and lesser part of the Gallatin plan was largely carried out during the period of active construction. Three of the four canals needed to provide an intercoastal waterway were built, including the Delaware and Raritan, though the fourth neck of land—Cape Cod—was not cut through until the twentieth century. Of the other canal construction in the East, a considerable proportion consisted of feeder or connecting lines related to the principal systems. One group of independent canals performed the particularly useful function of carrying to market the growing output of the anthracite mines.

Finally, there was a vigorous program of canal building in the western states, stimulated largely by the completion of the Erie and intended to extend the trade routes opened by that development. Here the most important projects were those connecting the Great Lakes with the Ohio and Mississippi rivers. Two major canals, with various branches, were built across the state of Ohio. The Wabash and Erie Canal connected the Lake with Indiana's principal river and became the nation's longest canal as it was extended to the Ohio River. The Illinois and Michigan Canal, making use of an even more obvious natural portage, connected Lake Michigan with the Mississippi system and did much to lay the basis for the future greatness of Chicago. In addition, the early construction of the Louisville and Portland

Canal provided a passage around the major obstacle to navigation on the Ohio River, and the later project of the Saint Mary's Falls (Soo) Canal opened an outlet for what was to become the tremendous commerce of Lake Superior.

All of this construction, east and west, was completed by the outbreak of the Civil War. In the century that has since intervened, expenditure on canals has been almost entirely confined to the handful of water connections that still appeared to present advantages in a railroad age: by the state of New York in replacing the Erie by the modern Barge Canal, and by the federal government on the Illinois Waterway, the Intercoastal Waterway, and on the enlargement of the short Soo Canal that carries the world's largest tonnage. The dollar total of these later appropriations, though exceeding that of expenditures in the antebellum years, is dwarfed by the sums devoted to railroads and highways and aviation and represents only a negligible fraction of total national investment.

The present volume asks two sets of questions with respect to the canal movement in ante-bellum America. The first concerns the decisions to build the canals. The second concerns their impact on the economy.

The decisions themselves were legion. What projects were to be chosen and on what criteria? What techniques should be employed in construction? How were the canals to be financed, and by what agencies were they to be built? Could a particular project be carried out, in spite of the difficulties that have been described, by unaided private enterprise? If not, should it be done entirely at public expense or by some combination of private and governmental means? If public action was required, should it be carried on by the federal government, by the states, or by the cities?

In the American case there was also another issue which the British canal builders had not had to face. Their system was virtually complete before the railroad was invented, but the newer innovation made its appearance while the United States was still in its first major wave of canal construction. American planners had therefore to put their questions in comparative terms. What were the relative advantages of the two types of improvement? Under what conditions should railroads be built instead of canals?

The first part of this study, which is concerned with the process of decision making, selects for analysis a small number of significant cases, each drawn from the early decades of the canal enthusiasm. They involve three states and four projects or systems. The first is devoted to the great innovation of the Erie Canal. Chapter I asks how the bold conception of the "Grand Canal" to Lake Erie came to replace more limited objectives and why a failure of mixed enterprise was followed by the notable success of a state work. Chapter II describes the reaction of Pennsylvania to the New York achievement. The need to build a rival line in a vastly different terrain produced the first public debate in the United States on the relative merits of canals and railways. Why did Pennsylvania reject the railroad alternative and attempt, in spite of the mountain barrier, a direct imitation of the New York project? These cases display a marked difference in the capacities of two states to devise and carry out successful programs of public works, and Chapter III presents the further contrast of a neighboring state which rejected both public and mixed enterprise. While New York was able to concentrate its initial effort on a single great project and Pennsylvania spent its energies on the simultaneous construction of a main line and various branches, New Jersey's interest was divided between two unrelated projects that commanded comparable amounts of public and political support. One of them, the Delaware and Raritan Canal to connect New York City with Philadelphia, appeared to offer such an opportunity for profit that Fulton was certain in 1796 that it would soon be snapped up by private capitalists.[20] Why, then, was it not constructed until a time when the greater share of the advantage would be won by a parallel railroad? The other, the Morris Canal across the hillier northern part of the state, was precisely the type of developmental enterprise that elsewhere was built as a public work or with heavy subsidy. How, then, was New Jersey able to obtain this improvement, as well as the other, without resorting either to public construction or to financial subsidy?

The second part of the study, which is concerned with the relations between the canals and the economy, employs a different method and examines the canal system as a whole. How large a part of the economic effort of the American people during the years 1815–60 was devoted to canal construction? What were its relations to their other economic activities? What effect, positive or negative, did it have on the economic growth of the country?

From Chapter IV, which is devoted to the process of construction, it appears that the canals were built not as the result of a single "canal mania" but in three long waves or "cycles." We must ask, therefore, how these fluctuations are to be explained. How was the timing of canal expenditure affected by the technical problems of construction, by the rivalries between competing states and projects, and by the ease or difficulty with which funds could be borrowed? How serious were the interruptions brought about by changes in the capital market? How were these canal cycles related to the more familiar business cycles affecting the economy as a whole? Did the waves of enthusiasm for canal building increase the instability of the American economy, or did expenditures on construction—sometimes extended into depression years—serve as a stabilizing influence?

Chapter V attempts to answer the still more difficult question of the impact of the canals on economic development. If their immediate effect was to cheapen costs of transport, by what chain of causation was it thought that this would lead to the opening of new areas of settlement, the creation of a wider national market, and a more efficient utilization of resources? What evidence is there that these effects did in fact occur? Were the benefits greater than the costs? The revenues of the canal system taken as a whole, in spite of certain financial successes like the Erie Canal, did not fully meet its costs, and the deficit became greater as the competition of the railroads became increasingly formidable. Yet neither contemporary observers, who spoke so often of the "incalculable" benefits of the canals, nor modern economists, who emphasize the importance of "social overhead capital" in development, could accept this criterion as wholly decisive. But if it is not, what other tests of benefit can be applied? Which canals were successes and which were failures, and what should be the judgment on the system as a whole? More broadly, what part did the canals play in a period of rapid economic growth characterized by the effective settlement of the Middle West and the rise of manufacturing industry in the East?

The book will end with this analysis of the economic consequences of the decisions to build the canals. But the decisions themselves, as we have seen, were largely those of governments, and we must first attempt to gain some understanding of the political processes by which they were made.

The Political Decisions

I. An Innovating Public Improvement: The Erie Canal

BY JULIUS RUBIN

The building of a canal between the Hudson and the Great Lakes in the 1820s captured the imagination of all Americans because it was the first breach in that great barrier to western economic development, the Appalachian Mountains. Like the first transcontinental railroad of half a century later, the canal proved to contemporaries that the methods of modern science and industry could overcome all such barriers to the expansion of the nation and thus provide the means to bind together the sectors of an enormous country. But the very success of the Erie Canal and the enthusiasm it aroused has distorted the writing of its history. Whereas much has been written of its drama and effects, there has been little concern with the causes of the 1817 decision to build it. Hindsight has distorted the nature of the decision: the effects of the Erie Canal have made its causes seem evident, the decision seem inevitable.

For those impressive effects flowed from a remarkable geographic situation. If ever there was a project that should have been built, whose advantages must have been obvious to anyone with the ability to read a map, it was that of a canal through the one gap in the Appalachians between Maine and Georgia; if ever there was a project that was obviously bound to succeed, it was that of an almost water-level canal that would link the Hudson to a chain of lakes that led directly into the heart of the continent. By means of a four-foot-deep, 353-mile ditch and seventy-seven locks, the northern Midwest could be converted into an island—as contemporaries put it—and New York City given access to one of the richest continental areas on earth. Hence there has seemed to be little in the decision to catch the historian's attention. Why not build such a canal?

Nevertheless, a closer look at the circumstances of the decision not only eliminates the obvious but arouses feelings of wonder and surmise. Contemporaries were not at all sure of the success of the project; for them, large-scale canal building was an entirely new and risky venture. Geography, far from determining the form the project finally took, actually seemed to dictate an entirely different, much easier, and less expensive alternative to the Erie route. Finally, the scale, the expense, and the broad developmental goals of the project in its final form seemed to require state construction; this meant that a political body responsible to a financially conservative and locally oriented electorate had to concentrate the energies and resources of the entire state upon a transportation line that would greatly benefit some sections of the state while it would not benefit or even would harm other sections. The 1817 decision to build the Erie Canal was therefore an unlikely choice among several alternatives; it was an investment of vast funds in a risky enterprise; and it was an example of centralized long-range planning in a democracy. The following pages will describe the long and complex history of the project in the attempt to discover the factors that made possible so difficult a decision.

Nature's Route to Lake Ontario

In its original form, the idea of a navigable communication between the Atlantic Ocean and the Lakes was not at all a leap into the unknown. Though it required a broad view of the geography of the continent, it required little of either imagination or daring to conclude from examination of a map that nature had presented New York with a gap through the Appalachians and a series of natural watercourses—the Hudson and Mohawk rivers, Wood Creek, Oneida Lake, and the Oneida, Oswego, and Seneca rivers—that could provide a continuous system of inland navigation from the ocean to Lake Ontario and Seneca Lake. Widen and deepen the rivers, remove the rocks and submerged timber, build short canals with locks across the portages and around the falls, and there would be the great navigable communication leading into the heart of the continent. Later, a short canal around Niagara Falls might connect Lake Ontario to Lake Erie. Without important innovations and at no great expense, New York City would have its line to the west. Nature having accomplished most of the work, it followed that a private company could finish the job, perhaps with some state assistance. Its simplicity accounts for the early origins of the idea; construction was delayed chiefly by the almost constant warfare on the New York frontier that ended only after the close of the Revolutionary War.

A notable 1724 memorial by the Surveyor-General of New York Province, Cadwallader Colden, began the century-long discussion that ended in construction of the Erie Canal. Colden's subject was the French domination of the fur trade of North America—a domination achieved by control of the one waterway that penetrated the continent. The St. Lawrence led into the Great Lakes; the Lakes led to the Mississippi River. With only a few short portages, canoes could carry freight from the Atlantic through the heart of the continent into the Gulf of Mexico. The French operated under serious disadvantages—the navigation of the St. Lawrence was slow and dangerous during most of the year and goods for the Indian trade were far cheaper at Albany than at Montreal—but as long as there was no alternative water route, French control of the fur trade was assured.

Colden insisted, however, that there was an alternative means of ingress into the continent, one with vast potentialities, for it also linked the Atlantic with the Great Lakes system. In use in Colden's time but barely developed, it began with that magnificent water highway, the Hudson, from New York City to Albany. From Albany, fur traders carried their Indian goods sixteen miles overland to the Mohawk at Schenectady; from there they could canoe up the Mohawk, Oneida Lake, and Oswego

River into Lake Ontario with but one long portage of three to five miles at what is now Rome. The distance between Albany and Lake Ontario by this route was about the same as the distance from Montreal to Lake Ontario, and the conditions of water transport were easier and safer. Why had it not been utilized on a large scale? Colden believed that the route had not developed because the decades of wars on the New York frontier had kept the Indians in a turmoil and had left Albany "exhausted." Now with peace, the New York merchants, who, he complained, had no regard for the interests of their country, found it easier to sell their goods to French traders from Montreal, who came to their doors, than to beat a path to the western Indians.[1]

The memoir contains a passing reference to the possibility of a route direct to Lake Erie which depended on the assumption that the Seneca River reached westward almost to the Lake. Colden withheld judgment at the time because so little was known of western New York. Later he undoubtedly abandoned the idea, for his *History of the Five Nations,* published in 1784, contained a map on which the Genesee River is shown quite accurately as running almost north and south between the Seneca River and Lake Erie. The factor that was for a long time to delay serious consideration of the direct Lake Erie route then became evident. As long as the necessity of following New York's natural watercourses was taken for granted, the route to Lake Ontario was the only conceivable one.[2]

Colden's perception of the immense potentialites of the Great Lakes—Mississippi complex and his realization that nature had provided New York with the key to this corridor into the continent were eventually to provide the conceptual bases of the great canal project. But the implementation of his vision was out of the question for some decades, for those wars that he had held responsible for the lack of development of New York's western route were shortly resumed and continued sporadically until the end of the Revolutionary War. Then, during the 1780s, the state's war debts precluded any substantial expenditures on suggested improvements.[3]

By the 1790s, however, the expansion of western New York and news of western road improvements in Pennsylvania gave the subject a new urgency, while the federal government's assumption of the state war debts and the sale of millions of acres of state lands placed New York in a position of relative affluence. In his January, 1791, message to the legislature, Governor George Clinton observed that the new conditions of peace and security on New York's frontier were producing a rapid increase of trade and settlement there and warned that the trade would be diverted to other markets unless the means of communication with the west were improved. There was no reference, however, to the Great Lakes system; the governor's view was entirely intrastate. Two months later the state's first canal law appropriated £100 to the commissioners of the land office for surveys along the Mohawk between Fort Stanwix and Wood Creek, for surveys between the Hudson and the Wood Creek that leads into Lake Champlain, and for estimates of the cost of making canals in these areas. The state's surveyors reported that the two canals would be inexpensive as well as practicable, and in February, 1792, a joint committee reported a bill for the incorporation of a private company to accomplish these tasks.[4] The state seemed ready to embark on this limited program for two small-scale, disparate improvements. But the bill proceeded no further; Cadwallader Colden's far-ranging vision, thus far dormant, was to become the basis for a radical reorientation of New York's policy.

The new perspective may be attributed above all to the vision and enthusiasm of Elkanah Watson. He was greatly influenced by the European canal experience, which he was able to observe at his leisure during five years of travel on the continent and in England during and after the Revolutionary War. Then in the winter of 1785 he spent two days at Mount Vernon in intense discussion with Washington of the latter's favorite project of connecting the Potomac with the Ohio. These experiences, Watson wrote, kindled "a canal flame in my mind," a flame which burned still brighter during three succeeding years of travel in the southern states, "at every step . . . contemplating the future benefit which would result from canals and improving the head branches of their rivers."⁵

The year 1788 found him at Fort Stanwix (now Rome) where, like Colden before him, he realized the immense significance of New York's route to Lake Ontario. On settling in Albany three years later, he became the first great promoter of the inland navigation project, able to communicate to his friends in the legislature an almost fanatical enthusiasm for a route that he considered nothing less than providential. "The various rivers, and their branches, lakes, and creeks," he wrote in 1791,

are disposed by the *Great Architect of the Universe* just as we would wish them. It is a circumstance as curious, as fortunate, that by opening a canal of about two miles at Fort Stanwix, a water communication of several thousand miles will be opened from the Atlantic to the most extensive inland navigation to be found in any other part of the world . . . the acquisition will be more precious than if we had encompassed the mines of Potosí.⁶

This broad view brought Watson into conflict with the above-mentioned bill for two short canals that was under consideration in the legislature in February, 1792. "I observe with regret," he wrote to General Philip Schuyler at that time, "that no one of that body [the legislature] (not even the Governor) appears to soar beyond Fort Stanwix except yourself. To stop at that point will be doing half the business . . . . The charter should stretch to the Seneca Lake, and to the harbor of Oswego . . . so as to admit the commerce of the great lakes into the Hudson River." To Watson, the crucial consideration in taking this view was time. New York State, he wrote, had every natural advantage. But without immediate attention to all the necessary improvements, "the channel of commerce may receive an early bias to a different point: and commerce is of such a nature, that when once established in any particular direction, it is generally found difficult to divert it." He therefore opposed the "European mode of calculation" which assumed that the West must be mature before canals were built. If, "calculating on the more enlarged American scale," canals preceded settlement, then "a vast wilderness will, as it were by magic, rise into instant cultivation." ⁷

Consequently, Watson drew up his own plan. In September, 1791, while the state was surveying for its two projects, he had traveled from Albany to Seneca Lake. On his return he gave to General Philip Schuyler, a prominent member of the state Senate, a report that traced the route of all the canals and described the improvements of all the natural water courses that were required to establish a navigable communication from the Hudson to Lake Ontario and Seneca Lake. It called for a canal to connect the Mohawk and the Hudson, clearing and some deepening of the Mohawk, a canal around Little Falls, a canal at Fort Stanwix to connect the Mohawk with Wood Creek, removal of obstructions and shortening of Wood Creek, improvement of the Seneca and Onandaga rivers, and canals at Seneca Falls and at Oswego Falls.[8] General Schuyler led the attack on the joint committee's bill for two short canals, managed to swing the Senate to his views, and was instructed to draw up a bill of larger scope.[9] Thus began the "lake canal policy" besides which the "little scheme" of connecting the Mohawk River and Wood Creek "nearly shrinks into nameless insignificance." [10]

The act of March, 1792, incorporated two companies: the Western Inland Lock Navigation Company was to open navigation from the Hudson River to Lake Ontario and Seneca Lake; the Northern Inland Lock Navigation Company was to connect the Hudson and Lake Champlain. Each company was to sell a maximum of 1,000 shares for an initial payment of $25 a share. As the work progressed, more money could be raised from assessments on the shareholders and from a state gift of $12,500 promised to each company after it had spent $25,000. The rate of toll was limited to $25 per ton from the Hudson to Seneca Lake or Lake Ontario; $20 per ton from the Hudson to Lake Champlain. The companies were granted fifteen years to complete the work.[11]

Financial difficulties plagued the Western Company almost from the time it opened its books in May, 1792. Sales of the stock were slow: three months elapsed before it sold 722 shares; three years passed before it sold twenty-one more. The company began with a capital of only $18,000, but in September, 1792, a survey committee estimated the cost of completion of navigation from Schenectady to Wood Creek, including canals at Little Falls and at Fort Stanwix, at £39,500 ($98,750); and in May, 1795, when the companies finally obtained an engineer (the Englishman William Weston), he estimated that the cost of providing navigation from the western extremity of Cayuga Lake to the Hudson at Troy would come to £189,497 ($473,742.50)—and this did not include the cost of a canal already completed at Little Falls. Consequently the company had to make frequent calls for further installments from the shareholders, who reacted by forfeiting a total of 240 shares between 1792 and 1795.

The company's last recourse was the State Legislature. The work, which began at Little Falls in April, 1793, was stopped in September for lack of funds, and it was feebly resumed in the following January only in the hope that some progress would induce the Legislature to come to the rescue. The hope was not misplaced; in addition to the gift of $12,500 promised in the charter, in March, 1795, the state subscribed to 200 of the shares of each company at the book value of $50 a share and in April, 1796, loaned the Western £15,000 ($37,500). The work then proceeded well. By 1798, the company had eliminated two portages along the Mohawk by means of canals at Little Falls and at Wolf Rift in German Flats and had completed a canal at Rome which linked the Mohawk to Wood Creek and the lake and river system beyond. The canals varied between nine tenths of a mile and almost two miles in length and were equipped with a total of nine locks. Periodic blasting and dragging operations were undertaken to clear the channels of the Mohawk and Wood Creek, and the creek was locked, straightened, and dammed.

The company began to collect tolls at Little Falls in 1796 and at Rome in the following year, but its financial difficulties continued. Public opinion in the west reacted sharply to the high tolls despite the fact that the company paid a first dividend of only three per cent in 1798, then could pay no dividends at all until 1813 when military traffic temporarily lifted revenues. Another appeal to the Legislature failed—no further aid was ever granted—with the important result that the company was unable ever to link its improvements to either the Hudson or to Lake Ontario. Finally, serious technical errors were made. Until 1795, the company's sole contact with an engineer consisted of correspondence with William Weston, then employed in Pennsylvania. Despite Weston's warnings, the locks of the company's three canals were built of wood, which rotted away within six years; on two of the canals these were replaced by brick and mortar locks which also failed; finally, under Weston's direction, they were all rebuilt with stone.[12]

For all these reasons, the Western Company was a financial failure. But this was not the important criterion. Nathan Miller has pointed out that to a large number of the stockholders the prospect of dividends must have been but a minor consideration. These merchants, bankers, and directors of insurance companies were land speculators "to whom the broad, unsettled stretches of land in western New York seemed like an untapped gold mine." Fifteen of the thirty-six known directors held lands which were likely to increase in value as a result of the company's improvements; one of them advertised the fact that lands he had put up for sale were accessible to the company's system of navigation.[13]

For many of the shareholders and directors, then, the most important objective was the lowering of transportation costs so as to increase property values. But in this respect too, the company's success was at least doubtful. For the company was unable to meet the competition of the turnpikes, which were superior in rates, speed, safety, and ease of transfer. A traveler along the Mohawk in 1809, surprised at the scarcity of boats, was told that carriage of freight by land was secure and punctual, but that the sudden and intense floods alternating with severe droughts made river travel extremely insecure. As a result, total toll revenues by 1803 amounted to less than $10,000 compared to $400,000 spent on construction and maintenance. In 1808 the company found that a toll reduction, initiated in response to rate-cutting by the wagonners, was not justified by the increase of traffic.[14]

It was evident that improvement of the natural watercourses did not produce a transportation route that was clearly superior to land carriage. This was a fundamental problem: it called into question the technical method that had been taken for granted since the time of Colden's memoir. It might be argued of course that the method had not been sufficiently tested, for with its inadequate financing, the Western Company could not eliminate the portage between the Hudson and the Mohawk in the east, and in 1808 it resigned its legal right to undertake improvements in navigation west of Lake Oneida.[15] Thus, the company that owed its origin to Watson's vision of an ocean-to-Lakes navigation was forced to limit itself to essentially local improvements. Nevertheless, the crucial limitations were those of the technical method. If the company's improvements had created a greatly superior route along the Mohawk, its high tolls would not have driven the trade to the roads. The small value of those improvements is indicated by the remarkably bitter reaction to the tolls. The company's collectors were subjected to "the grossest abuse" by boatmen and migrating settlers, and merchants of Geneva and Canandaigua circulated petitions which denounced the tolls as "oppressive" despite the fact that the shareholders received almost no return for twenty years.[16]

In every respect, then, the Western Company was a failure. That failure had important consequences. Many became sceptical of all such projects; indeed, De Witt Clinton later charged that the Western Company experience gave rise to the most popular and the strongest arguments against the Erie Canal policy.[17] But for others the lessons of the failure were to pave the way for an ultimately successful reorientation. The chronic shortage of the company's funds forced the state's improvement advocates to the realization that they were dealing with an immense and complex project that required long-range state financing and possibly federal aid. The failure of private enterprise and the public reaction to the Western Company's tolls, when taken together with the great sacrifices which would be required, created a strong bias in favor of state construction.[18] Finally, the failure of river improvement led to the consideration of a large-scale innovation—a full canal separate from the natural watercourses—and that raised the possibility of a route entirely different in direction from that of the watercourses of the state. In sum, the failure of the Western

Company helped to pry some men loose from their accustomed preference for the simplest and cheapest possible modification of nature; by eliminating this alternative it forced them to think of large-scale expenditures, long-range public planning and construction, and the radical transformation—not the mere improvement—of the state's geography.

Man's Canal to Lake Erie

Neither the concept of the full canal independent of natural watercourses nor the idea that it might go directly to Lake Erie was at all difficult to conceive in theory: many did so at an early date. Much more difficult and rare, however, was to take seriously the political and economic feasibility of either idea. As early as 1777 Benjamin Franklin had reported from London that Europeans were fully convinced of the great advantages of canals that were separate from natural watercourses:

> Here . . . they seldom or ever use a River where it can be avoided. . . . Rivers are ungovernable things, especially in Hilly Countries. Canals are quiet and very manageable. Therefore they are often carried on here by the Sides of Rivers . . . no other Use being made of the Rivers than to supply occasionally the waste of water in the Canals.[19]

But in America these advantages seemed minor compared to the cost and technical difficulty involved. In an 1808 letter to Albert Gallatin, for example, Benjamin Latrobe complained of the American tendency to think of canals only as a last recourse, to be used only where navigation was otherwise impossible. He thought that it was the scarcity of engineers and of capital, and "above all the erroneous dread of bold measures, and the fear of uselessly expending money in works hitherto unknown to us" that had prevented Americans from deserting the beds of their rivers. But if the English preferred full canals, Latrobe remarked, how much more should the Americans, whose rivers were subject to much more severe floods and droughts.[20]

Similarly the idea of the Erie route had appeared at an early date, but its feasibility was long discounted. The idea was discussed as early as 1777 in conversations at Little Falls between the president of the Western Company, General Philip Schuyler, and the engineer, William Weston. They agreed that an "interior" route parallel to Lake Ontario and avoiding Niagara Falls might be technically possible—though little was known of the western country at the time—and would have the important advantage of by-passing the Canadians on Lake Ontario; and they were well aware of the advantages of the full canal over river improvements. But, in view of the "infant state of the country" and public opinion, "they considered the period remote when this great system of canalling was to be adopted."[21]

In this period, then, the Erie Canal project could be taken seriously only by a visionary, a man so long-sighted that he could minimize, ignore, or simply be unaware of the host of practical difficulties in the way. Gouverneur Morris was such a man. He had Elkanah Watson's quality of flaming enthusiasm but combined it with a contempt for mere facts—an attitude which Watson was incapable of taking. During the darkest period of the Revolutionary War, in 1777, he spent his evenings at Fort Edwards describing in glowing terms to his comrades-in-arms "the rising glories of the western world," and "the rapid march of the useful arts through our country, when once freed from a foreign yoke." One evening, he predicted "in language highly poetic" the day not far distant when "the waters of the great western inland seas, would, by the aid of man, break through their barriers and mingle with those of the Hudson."[22]

Morris was deeply involved in national politics until 1789, then spent almost a decade in Europe. The year 1800, however, found him traveling in western New York, where he experienced a vision similar to that of Elkanah Watson twelve years before. On turning a point of the woods he came suddenly upon Lake Erie and saw riding at anchor nine large vessels. Inspired, he wrote a prophetic letter to a friend:

Can you bring your imagination to realize this scene? Does it not seem like magic? Yet this magic is but the early effort of victorious industry. Hundreds of large ships will in no distant period bound on the billows of those inland seas. At this point commences a navigation of more than a thousand miles. Shall I lead your astonishment up to the verge of incredulity? I will: know then, that one-tenth of the expense borne by Britain in the last campaign, would enable ships to sail from London through Hudson's River into Lake Erie. As yet, my friend, we only crawl along the outer shell of our country. The interior excels the part we inhabit in soil, in climate, in everything. The proudest empire in Europe is but a bubble compared to what America *will* be, *must* be, in the course of two centuries, perhaps of one.[23]

The source of this unbounded optimism was not only the prospect of western expansion. It was the speed of the recent development of the country that made the most prodigious projects seem possible—at least to men like Morris. In that same letter a conservative projection of the course of federal revenues and population led him to predict that in 1840, £30 million would be raised from a population of thirty millions. "This, indeed, seems impossible," he admitted, "but did it not seem equally impossible at the close of the Seven Years' War that the net revenue of British America should exceed two millions sterling by the end of the century? Had this been asserted on the Exchange of London in the year 1760, would it not have been laughed at?"

Three years later Morris evidently mentioned to Simeon De Witt the project of "tapping Lake Erie" and of leading its waters in an artificial river across the country to the Hudson. To the objection that there were hills and valleys in the way, Morris answered, characteristically, that the goal would justify any amount of labor and expense. At the time, De Witt considered the scheme "a romantic thing, and characteristic of the man," and this was the judgment of his contemporaries. But one factor in the Erie Canal idea did impress a practical mind like that of James Geddes. By the Ontario route, one canal would have to descend from Oneida Lake to Lake Ontario; a second, around Niagara Falls, would have to ascend to Lake Erie. A canal rising directly to Lake Erie would therefore save twice the lockage between Oneida Lake and Lake Ontario. Hence, when Geddes heard of the idea from De Witt in 1804, he concluded that it was the saving of lockage that had probably suggested the route to Morris. Nevertheless Geddes was unconvinced. He urged Morris not to ignore "the ready made navigation" to Lake Ontario, but "such was his [Morris's] ignorance of the country to be passed, and his pertinacity was such, that it was almost impossible to call his attention to the impracticability of such a thing." [24]

Morris's enthusiasm undoubtedly kept the canal idea alive in a period of disillusionment with improvement projects in general. But the very qualities that allowed him to do this—a dislike of detail, a cavalier attitude towards difficulties, an aristocratic independence that often reached the point of irascibility —prevented him from winning an influential group to his ideas. Furthermore, the resistance to the direct Erie route must have been enormous. No one, not even Morris, had yet defended its practicability in the public prints, despite its great advantages. After all, the route that seemed to be dictated by nature, the one that allowed the use of the simplest and cheapest of methods—river improvement plus short canals—had been the focus of interest for over eight decades. If all the discussion and efforts of those years had not produced the money and enthusiasm for the completion of that project, it may well have seemed a waste of time to try to defend publicly a route that would require a major innovation and would be twice as long because it would run parallel to that gift of nature, Lake Ontario. Hence, as Elkanah Watson testified, "From the conception of Morris in the year 1800, to the year 1808, nothing was done. . . . The community stood appalled in contemplating the vast enterprize; but none dared approach it in earnest. It was occasionally the subject of conversation, generally ending in ridicule." [25]

It is all the more remarkable, then, that in 1808 the Erie Canal idea did come to the fore and was favorably acted upon in the legislature. The first public exposition of its advantages appeared in a series of essays signed "Hercules," published by the Canandaigua *Genesee Messenger* between October 1807 and April 1808. (An introductory essay had appeared in the January 14, 1807 issue of the Pittsburgh *Commonwealth*.) The author was Jesse Hawley, an obscure merchant with little formal education, who had never read a treatise on canalling, except for some short sketches in geographies and encyclopedias, and had never even seen most of the areas he wrote of. In 1805 he was engaged in the flour trade at Geneva. Often irritated by the difficulties of navigation on the Mohawk, he and his fellow merchants would express the wish, as he later recalled, "that an arm of the North River had been extended into the Genesee country by the Author of nature, for our facilities and transport." One day it occurred to him that man could make up for the oversight. He suggested to his friends that the waters of Lake Erie might be carried by canal from Buffalo to the Mohawk at Utica—and was thoroughly ridiculed for the idea. Soon after, a business failure that placed him in danger of a debtors' jail forced him to flee Geneva. Arriving at Canandaigua in August, 1807, he determined to publish his idea: "Fully persuaded of the practicability of such canal; and having, thus far, lived to but little purpose, I thought I might render myself useful to society by giving publicity to the suggestion." This was the origin of the Hercules papers.[26]

Hawley's plan was for a 200-mile "Genesee Canal," 100 feet wide and 10 feet deep—the dimensions of France's Languedoc Canal—that would carry the inexhaustible supply of Lake Erie water down to the Mohawk at Utica by a route that was almost the one later taken by the Erie Canal. Since the canal would be carried along an inclined plane, falling 410 feet as it ran eastward, it would require as few as twenty-six locks. Much of it, he thought, would require simply the digging of a ditch and the finding of some local stones for the banks. The immensely increased supply of water on the Mohawk, together with the deepening and other improvements on that river, would provide a good navigation to Schenectady and later, when that last portage was eliminated by a canal, to the Hudson. The project was to be carried through by the federal government.

This was the entire proposal and principal argument; there were no technical details. The only additional defense of his project consisted of references to the success of the European canals. "Pretending to no knowledge in the science of canalling," he could estimate the expense only hesitantly at $6 million, on an analogy with the Languedoc and the Clyde canals. "Having never seen any part of the route spoken of, but the villages at the two extremes," he admitted that he might be "minutely incorrect in some particulars." But he could claim that his only purpose was to justify a survey. Throughout he kept his eye on the main point, which was that the "parent of nature" had not neglected the west, as many complained, but had actually laid the basis for supplying the region with as magnificent a waterway as that possessed by eastern New York. "By the Falls of Niagara he [the Creator] has given a head to the waters of Lake Erie sufficient to flow into the Atlantic by the channels of the Mohawk and the Hudson, as well as by that of the St. Lawrence. He has only left the finishing stroke to be applied by the hand of art, and it is complete! Who can reasonably complain?"

Ill-informed, extreme in tone, bady organized, and poorly expressed, Hawley's plan was nevertheless the first of the great New York transportation visions to combine so many of the elements of the future successful program: government planning and construction, the canal method, and the Erie route. Two technical aspects would be changed by his successors: they would choose a canal that followed the contours of the land by means of locks rather than alternately dig and build up an inclined plane; and they would prefer complete separation from the watercourses to dependence upon the vagaries of the Mohawk. But these were minor aspects. The important thing was that the Erie route was at long last before the public for discussion.

At the very time that the *Genesee Messenger* was publishing the Hercules papers, the canal issue became the subject of intense public discussion in Onondaga County (of which the largest city is Syracuse), and a Federalist, Joshua Forman, united with a Democrat, John McWhorter, to run successfully for the Assembly on a "canal ticket." [27] The following February, Forman and Benjamin Wright of Oneida County proposed to the Assembly the formation of a joint committee to consider a survey of the "most eligible and direct" route for a canal between Lake Erie and the Hudson. Forman estimated its cost at $10 million by taking the $4.5-million cost of the famous Languedoc Canal of France, multiplying by two to take account of the high price of labor in the United States, and adding a million for "inexperience." The resolution passed, though it encountered a good deal of ridicule, and a joint committee was appointed. The committee, obviously interested in the direct route to Lake Erie but afraid to put it in so many words, ordered the Surveyor-General "to cause an accurate survey to be made of the rivers, streams and waters . . . in the usual route of communication between the Hudson River and Lake Erie, and . . . such other routes as he may deem proper." The totally inadequate sum of $600 was appropriated.[28]

These events of 1808 were of decisive importance because they led to the survey that established the practicability of the Erie route. But this achievement was almost inadvertent, so far as the state was concerned. With only $600 appropriated, the Surveyor-General, Simeon De Witt, ordered James Geddes to survey only the areas between Oneida Lake and Lake Ontario and between Lake Ontario and Lake Erie. As for the direct route to Lake Erie, De Witt relied in part on information supplied by Joseph Ellicott, the resident agent of the Holland Company, which held a huge tract of land in western New York for a group of Dutch investors. Ellicott had sent De Witt a general description and possible canal route for the country between the Tonnewanta Creek (near Buffalo) and the Genesee River. Geddes was therefore instructed to explore only the area between the Genesee and the Seneca rivers, but without leveling or surveying, as the appropriation would probably have been insufficient for those purposes.[29]

Geddes nevertheless did a crucial bit of surveying in this area. He agreed with Ellicott—since 1805 he had believed that the area west of the Genesee was suitable for a canal—and he had little doubt concerning the country east of Mud Creek (at Palmyra). But the area between the Genesee and Mud Creek might well present the great obstacle to an Erie route, for it was unknown and was generally assumed to contain high land with no possible source of water. He set to work there in December, 1808, using his own money because the state's appropriation had been used up on the other route, and soon found—at a cost of

$73—not only that the Genesee was far above that high land, but that that river water could be brought across the intervening Irondequoit valley to Mud Creek along natural ridges that in many places were just sufficient in height and width to support a canal. With no upward lockage required from the Genesee to Cayuga Lake, the probable practicability of the Erie route had been demonstrated. But even these results—far more favorable than he had dared to expect—did not produce in Geddes a belief in the actual construction of the canal during his lifetime. Posterity, he thought, would one day "enjoy the sublime spectacle, but . . . for myself, I had been born many, very many years too soon." [30]

With the Geddes report, the Erie route took its place as a practical alternative to the Ontario route and a serious assessment of the relative advantages of the two was begun. Geddes listed the arguments in his report but took no position. Against the Ontario route was the fact that the products of the upper lakes, once brought onto Lake Ontario by a canal around Niagara Falls, would be confronted with the choice of 206 feet of lockage to tidewater on the St. Lawrence as against 574 feet of lockage on the New York canal to Albany. It was possible, Geddes warned, that the entire lake trade might be diverted to a foreign power. In addition, the Erie route would be safe in case of a war with Great Britain, it was not subject to the open weather of a lake, and it would raise land values and increase population in a much larger area of western New York. In favor of the Ontario route, on the other hand, was its "cheapness of conveyance, the grand desideratum in all such works," for its port on Lake Ontario at Oswego would be 150 miles closer to the Hudson by canal than would be the port of Black Rock on Lake Erie. However, Geddes raised the possibility that the cost of construction would be about the same on the two routes. In that case western New York beyond Oswego would obtain from the Erie alternative a very great reduction of its transport costs at no extra expense.[31]

What factors explain the sudden rise of the Erie alternative in 1808? The writing of the Hercules papers and the results of the Geddes survey may be attributed to individual talents or peculiarities. But the winning "canal ticket" in Onondaga County, and particularly the success of the Forman resolution in the face of legislative ridicule, indicate a widespread change of attitude towards a project that might well have seemed too stupendous for consideration. Two interrelated trends provide a good part of the answer. One was the expansion of the west; the other was the national movement for a federal transportation plan.

Westward expansion was particularly rapid in New York because the natural route along the Mohawk valley had been blocked for so long by the most powerful Indian force on the continent. New York in 1749 had been one of the less important colonies, with a population of only 73,448. From the mid-eighteenth century, however, the rate of growth was explosive: 168,007 in 1771, 340,120 in 1790, 959,049 in 1810. A number of factors combined to bring the state to the forefront in so short a period: the pushing back of the Indians together with the elimination of the French menace, the great fertility of the Mohawk valley, and the seemingly endless stream of immigrants from New England. But the settlement of western New York did not proceed at a uniform pace. The leap up the Hudson valley was succeeded by a much slower movement along the Mohawk, for the Indians continued in strength in the Genesee area, and the British retained their lake ports until 1796. In 1790 the census showed only 1,075 people west of Seneca lake; in 1800 there were still only 17,016, and settlement scarcely extended to the Genesee River. By 1810, however, the number had risen to 72,000 and in 1820 to 211,000.[32]

Since there was relatively little need for improved transportation as long as the population was expanding along the Hudson valley, intense western interest in the canal project would be expected to appear only toward the end of the eighteenth century. Indeed, it was in 1791 that Governor George Clinton warned the legislature that the rapid expansion of the west made an improvement policy urgent. And only with large-scale settlement south of Lake Ontario in the nineteenth century would we expect an important western interest in a route direct to Lake Erie. This is the factor that explains the emergence of local western figures in the crucial events of 1807–9. Previously it was the great men of imperial or national views, like Cadwallader Colden and Gouverneur Morris, who raised the issue of an ocean to lakes policy; but in 1807–9 it was the Geneva merchant Jesse Hawley and the Onondaga county "canal ticket" politicians, Joshua Forman and John McWhorter, who came to the fore and transferred the idea to the realm of the practical.

The success of that canal ticket as early as 1807 indicates the ease with which western support could be obtained. For a westerner, in distinction to a resident of New York City, did not have to be convinced of the practicability of the entire immense project; local interest was enough. According to the recollections of Thomas Wheeler, then of Salina, Forman tried to obtain his support for the Assembly by persuading him of the practicability of an Erie Canal. "Thus he urged me for more than an hour; but I refused . . . and then I thought, if a canal was only made from Rome to Cayuga Lake, it would be of great advantage to this section of the State, and so I agreed to support him." [33]

The other major factor in that sudden change of attitudes after 1807 was the national movement for federal aid to internal improvements, itself the result of the expansion of the west. This movement got its start in 1806 when Jefferson suggested that the Treasury surplus, which was expected soon if peace continued, might well be devoted to public education and internal improvements. Later he repeated the idea when he noted that the federal surplus was even greater than anticipated. Then the lively Congressional discussions of aid to specific projects was followed by the Gallatin *Report* with its proposal that the federal government spend $20 million of its surplus during the next ten years on a national transportation network.[34]

The *Report* attributed the undeveloped state of American transportation, on the one hand to the scarcity of capital which made it unavailable for large projects, on the other hand to the sparseness of the population which made large projects the only profitable ones. For there were few canals that could be undertaken with a view solely to the trade between the two extremes and along the route of the canal itself; to be profitable, the canal had to connect with a large natural body of water, thereby making up in extensiveness for the low density of population. Hence private construction of improvements was generally unprofitable, for these small projects depended for their full potential upon other improvements too extensive or too distant to be built by the same company. "The general government can alone remove these obstacles."

The principal large-scale projects proposed in the *Report* would have spanned the country—a tidewater navigation and a turnpike from New England to Georgia; and from tidewater five routes to the west. But four of these east-west routes would have had to climb the Appalachians; on these a canal was technically impossible, and Gallatin had to recommend roads between the western and the improved eastern rivers. New York's route, however, turned the mountains: the Hudson and Mohawk river valleys provided the only possibility of a through navigation to the west. The *Report* called for the expenditure of $4,000,000 in New York State: $2,200,000 on a canal from Troy to the eastern end of Lake Oneida and from the western end of that lake to Lake Ontario; $1,000,000 on a canal from Lake Ontario to Lake Erie; and $800,000 on a canal, river, and lake navigation between the Hudson and Lake Champlain. There was no reference to the possibility of a route direct to Lake Erie.[35]

The Gallatin *Report* was issued precisely at the time that the Hercules papers were appearing in western New York. It is of interest to compare them, for, although the Geneva merchant's views were far more extreme and his information far less precise, he too called for public enterprise and national plans. Like Gallatin, Hawley considered the supply of domestic capital inadequate to large projects. But unlike Gallatin, he objected in any case to private enterprise in transportation. The import of capital to make up for the domestic shortage particularly disturbed him. Unless the government intended an ultimate "sequestration of capital," he declared, there could be no excuse for allowing foreigners to draw "by their superior wealth, a private revenue from our best resources." But he opposed domestic as well as foreign private capital, for both levied tolls and both displayed "that prejudicial propensity to which incorporate bodies are subject, in their divergings from the common interest, their tendency to monopolies." This attitude was not at all a peculiarity of Hawley's; it was held with remarkable unanimity by the early advocates of a lakes-to-ocean project. In 1791, before the experience of the Western Company had demonstrated the inadequacy of private capital, Elkanah Watson had suggested that his project might be built either by the state government or by private enterprise. But the wisest policy, in his opinion, was state construction in order that the passage might be left free, "as otherwise posterity will be burthened with a weight, (in the article of toll,) to the emolument of the successors of the first adventurers—which ought not to exist in a land of liberty; where the intercourse should be as free as the air in which we breathe." [36]

For Hawley, then, the government was both the necessary and the preferred agent of construction. But by "government" he always meant the federal government. He began his articles with a reference to Jefferson's proposals and not once did he mention the possibility of construction by the state. For in his view, as in Gallatin's, the effects of the New York canal would be national in scope; it would help bind the sections of the nation economically, politically, and even socially by helping "to assimilate their manners in the infancy of our country." Furthermore, the national government would recover its expenditures in the increased customs revenues made possible by the canal. But again he went much further then Gallatin in calling for an amendment to the Constitution that would enable the federal government to apply the surplus revenues to internal improvements and to exercise "the uncontrollable right to enter, range through, and leave the territory of any individual state at discretion." This would eliminate the impediments of local prejudices and selfish jealousies—"leaving old places to flourish or decline, and new ones to arise, as natural advantages decided."

Hawley went on to describe the national policy that would then be possible. There were first of all, the possible extensions of his Genesee Canal, for the Creator had supplied not only Lake Erie but all the other Great Lakes as well as the Mississippi and the Ohio. Hawley listed a series of small projects—the improvement of the straits of St. Marie to complete the navigation into Lake Superior, a canal and river improvement to connect Lake Michigan with the Mississippi, and many others—and could then trace the route of a boat from New York City to the Gulf of Mexico. By means of that one great finishing stroke and these few small improvements, New York would stand in competition with New Orleans for the trade of the entire northwest.

But Hercules did not stop even there. "We entertain vast ideas of the destinies of these United States," he declared, and proceeded to document his case. An intensive study of geographies and gazetteers had convinced him that nature had distributed her favors not only to New York but to all of America "with . . . a peculiar magnificence; and nature invites Americans to project their plans of internal improvements on her magnificent scale." He saw the possibility of a multitude of improvements in every part of the continent and—with a warning to his readers that his information was "general, not particular," that he knew no engineering, and that therefore he might be mistaken in some cases—he listed them all, described their immense benefits, and ended with the "most noble work of its kind," a canal through the Isthmus of Darien.

Here, then, were two national plans, one by a local New York merchant, the other by the Secretary of the Treasury of the United States. Both were inspired by the movement for a national transportation policy. But if the virtue of Hawley's contribution was that it set before the public an original and radical project, the virtue of Gallatin's *Report* was that it made such projects respectable. Under the sponsorship of Jefferson himself and in a tone that was sober, precise, and practical, it transmitted to a national audience the breadth of vision of men like Colden, Watson, Morris, and Hawley. Furthermore, it calculated exactly the means that had been promised more vaguely by Jefferson: those large concrete figures must have vastly increased the general confidence in the imminence of the federal subsidy. It was this expectation that set things moving in New York, for, if the immense cost was to be shouldered by the federal government, the major objection to serious consideration of projects such as Hawley's was removed. Finally, the *Report,* and the series of events of which it was a part, vastly intensified the competition of the seaboard cities for the western market. For the imminence of federal aid in specific amounts to particular projects goaded the states to immediate action—to surveys, cost estimates, and plans that would maintain or increase the share of each against the other. This rivalry, too, tended to reduce the importance of narrowly conceived cost and profit calculations. As Hugh Williamson put it in 1807,

This western country may be compared to a handsome girl, who has two rival lovers—the one, Baltimore, flattering her fancy—the other, New York, too sure of conquest, and therefore neglectful of his courtesy. . . . If you wish New York to remain, as now, the emporium of America, suffer not the trade with the interior of your state to be carried off triumphantly by the spirited and enterprising citizens of Philadelphia and Baltimore.[37]

Once fixed by the spirited enterprise of rivals, these new trade connections would be extremely difficult to alter. A group of Ohioans therefore warned the New York Canal Commissioners in 1810 that postponement of a project because of inadequate prospects of immediate profit might well result in the permanent loss of a trading area.[38]

These decisive effects of the national movement are particularly apparent in the history of the Forman resolution. According to Benjamin Wright, Forman's idea germinated when he read the article on "Canals" in Rees's *Cyclopedia,* then came across a newspaper account of a recommendation by a committee of the Pennsylvania legislature for state construction of a road from Philadelphia to the Genesee area. "Upon reading this, Judge Forman observed that something ought to be done to prevent the people of Pennsylvania from drawing away the trade of our state, and suggested that as Mr. Jefferson . . . had recommended the surplus monies of the treasury to be expended on roads and canals, he was for making a canal to Lake Erie." [39] Consequently Forman began his resolution with a reference to Jefferson's suggestion, then referred to the great exertions of other states to secure the trade of the west "under natural conditions vastly inferior to those of this state," and ended with a resolve for an Erie to Hudson survey "to the end that Congress may be enabled to appropriate such sums as may be necessary." [40] Thus did those federal funds—which were destined never to materialize—play a crucial role in the determination of the practicability of the Erie Canal.

The Project Proposed

The result of these developments of the 1807–9 period was the abandonment of the state's traditional improvement policy. The new turn came in March of 1810, when Thomas Eddy, a director of the Western Company, suggested to State Senator Jonas Platt, who represented the Western District, that the legislature might appoint a commission on the possible westward expansion of navigation from Oneida Lake to the Seneca River by means of a canal to be built by the Western Company. Platt refused his support; instead he proposed a plan for a full canal from the Hudson to the Lakes independent of the watercourses and untarnished by the disappointed hopes associated with the name of the Western Company. No private corporation, Platt declared, was adequate to or could be entrusted with power and control over so important an object. Eddy objected that the Legislature would be so frightened by the magnitude of such a proposal that nothing at all would be obtained. But Platt saw a way out: if De Witt Clinton, one of the most influential figures in the state Republican organization, could be persuaded to join forces with Platt, who was the Federalist minority leader of the Senate, the bill would be carried.

The next day Platt approached Clinton with a carefully worded resolution that left implicit his full proposal: whereas the agricultural and commercial interests of the state required improvement of the navigation from the Hudson to Lakes Erie and Ontario, and whereas it was doubtful whether the resources of the Western Company were adequate to such an improvement, his resolution proposed that commissioners be appointed to explore the entire route and to recommend a program to the next session of the Legislature. Clinton remarked that he had given little attention to canal navigation, but the subject appealed to him. He agreed to second the resolution. It passed both Senate and Assembly unanimously, and $3,000 was appropriated for the commissioners' expenses. The unanimity is not surprising. For the proposal was as yet exploratory, not specific; and in the midst of the national improvement movement a group of state-wide political leaders had taken over direction of the project. At least as far as exploration was concerned, the fate of the great canal project was no longer in the hands of visionaries alone.

The commissioners were carefully chosen. There were an equal number of Federalists and Republicans, but all were men of "wealth and public spirit" who would demand no compensation. They were Gouverneur Morris, De Witt Clinton, Stephen Van Rensselaer, Simeon De Witt, William North, Thomas Eddy, and Peter B. Porter—men whose recommendations would certainly favor a full-scale project.[41]

The commissioners, aided by James Geddes as surveyor, explored during the summer of 1810, taking with them Geddes's 1809 report, Joseph Ellicott's letter and map of the country between the Niagara and Genesee rivers, and Jesse Hawley's essays. Their report, presented to the Legislature in March, 1811, marked the beginning of a new era: for the first time, a responsible official body recommended state construction of a full canal direct to Lake Erie.[42]

First, the policy of improving rivers was rejected; it was considered outmoded in Europe and proven inefficient in the United States by the experience of the Western Company.[43] Then, since the direction of the state's watercourses was no longer determinative, a canal direct to Lake Erie was conceivable. Consequently, the commissioners stressed for the first time the physical disadvantage of the Ontario route—the supply of water, inadequate for river and lake schooners and barely adequate for boats; the particularly difficult section from the falls in the Oswego River to Lake Ontario for which they suggested a horse railroad; the need for many transshipments on a route that would consist successively of a river, a canal, a railroad, a lake, then eventually a canal again between Lake Ontario and Lake Erie; finally, the specter of Canadian competition.

The commissioners therefore preferred the Erie route. But they rejected for it the generally accepted method of a lock canal fed by rivers. For the Erie route, they pointed out, had a further advantage over the Ontario route: it could be fed by lake water. It need merely be constructed on an inclined plane, as Jesse Hawley had suggested, with a uniform descent from Lake Erie of six inches per mile, maintained by means of huge aqueducts across valleys and considerable winding around hills. The plane would end at the ridge between Schenectady and Albany, from which the canal would descend three or four hundred feet by locks. Here and there the commissioners hedged—they repeatedly emphasized the need for a more detailed survey before coming to a decision as to method—nevertheless they stressed the great advantages of the inclined plane.

These advantages were quite impressive. There would never be a shortage of water even in the dryest season; indeed the canal could supply Lake Erie water to any area that needed it. With no locks on most of the canal, maintenance costs, and therefore tolls, would be minuscule. Moreover, a side cut of only five or six miles would, by means of locks, connect the canal with Lake Ontario at the harbor of the Genesee River. A similar connection could be made with both Seneca and Cayuga lakes, from the heads of which there was only a short portage to the Susquehanna River which in turn led to Chesapeake Bay. Thus a variety of markets would "stimulate and reward the industry" of the settlers on the fifty million acres of land that bordered the Great Lakes (exclusive of Lake Superior).

The cost of such a canal would depend upon its dimensions. In contrast to an Ontario canal, the Erie, with its unlimited supply of water, could be made large enough to accommodate the eight-foot-draft sloops of the Hudson and the Lakes. But the commissioners doubted the desirablity of doubling expenditures in order to avoid transshipments at the ends of so long a canal. For a boat canal, costs would be quite low. The nature of the land would make excavation easy; stone, limestone, fuel, and cheap subsistence for laborers would be plentiful all along the route; in addition, "good bargains for the public" could undoubtedly be negotiated with the owners of land on the route, and in fact many generous offers had already been made. With repeated warnings that further surveys were needed for more definite estimates, they put the cost of their inclined-plane boat canal at $5 million and were certain that "there is no part of the civilized world where an object of such great magnitude can be compassed at so small an expense." Even if the cost were $50 million, they declared, it would not exceed half the value of the goods that "at no distant period" would be carried along the canal.

Those millions, however, would have to come from Washington. Like Jesse Hawley and all the other prominent advocates of the grand canal up to that time, the commissioners assumed that private capital would be inadequate and federal subsidy essential. Had there been no federal ties, they threatened, New York might well have found it advantageous to build at its own expense, then levy a transit duty, "raising and lowering the impost, as circumstances might direct for her own advantage." But fortunately there was a federal government which the commissioners were confident would exercise "prudent munificence."

That the commissioners really expected massive federal aid is indicated by their reference to one of the huge aqueducts that would be required by their inclined-plane canal. To cross the mouth of Cayuga Lake would require a structure extending about a mile at an elevation of 130 feet: "to erect a mound of that length and of the sufficient height and breadth, is an herculean labor," the commissioners admitted. "Whether it will be performed, must depend on the arm that undertakes this task." The only Herculean arm around was that of the federal government. No such language would have been used, had the commissioners wished to prepare the state for the assumption of the burden. Clearly, independent state action was as yet only a threat to make certain of obtaining the federal subsidy.

Although the act of April, 1811, again did not specify the route, it indicated general legislative acceptance of the viewpoint of the commissioners. Stressing almost exclusively the advantages of the canal to the nation as a whole, it created a board consisting of the old commissioners, plus Robert R. Livingston and Robert Fulton, for the consideration of all matters relating to "canal navigation between the great lakes and Hudson's river" and appropriated $15,000 for its use. The board was empowered to make applications for aid from Congress and the legislatures of other states; to accept cessions of land from proprietors along the route of the proposed canal; to ascertain the terms on which loans could be procured; and to negotiate with the Western Company for the surrender of its rights.[44] Clearly the Legislature was willing to proceed so long as a large proportion of the expense would be paid by others.

But the expectation of federal subsidy, seemingly so justified by Jefferson's statements and Gallatin's *Report,* was in for a series of rude shocks. A preliminary tremor had occurred in January, 1809, when Joshua Forman and William Kirkpatrick, a Congressman from Sabina, visited Jefferson to inform him that, in view of his proposal to spend the federal surplus on roads and canals, New York had explored the route of a canal from the Hudson to Lake Erie and found it practical "beyond their most sanguine expectations." According to Forman's recollections, Jefferson replied that it was certainly a very fine project and "might be executed a century hence. 'Why sir,' said he, 'here is a canal of a few miles, projected by General Washington, which, if completed, would render this a fine commercial city, which has languished for many years because the small sum of $200,000 necessary to complete it, cannot be obtained from the general government, the state government, or from individuals—and you talk of making a canal of 350 miles through the wilderness' "[45]

Succeeding events in Congress were no more encouraging. On February 8, 1810, Representative Peter Buell Porter of western

New York delivered an impassioned speech for a national internal improvement plan to be financed by the appropriation of the public lands. His bill followed the Gallatin *Report* closely, both in advocating all the projects suggested there and in assuming the Ontario route for New York. But his financing scheme, he claimed, would involve no burden upon anyone. The federal government would issue certificates carrying six percent interest, each eventually redeemable out of the proceeds of a particular tract of land. These certificates would be sold on the market or given to the builders of the projects. The completion of the projects would provide markets and speed settlement, thereby immensely enhancing the value of the public lands. The success of the plan depended only on the rapid growth of the country, and this was certain in view of the unbelievably rapid growth of the recent past. The present time was particularly favorable, Porter declared, because "the great commercial capitals" and the "vast numbers of our sailors and other labourers" thrown out of work by the stagnation of foreign commerce might be engaged in these improvements.[46]

Congress, however, was unimpressed. A committee shortly reported the bill he advocated, identical with a bill which Senator John Pope of Kentucky had introduced in the Senate on January 5, 1810. But Porter's bill was read twice and referred to a committee of the whole which never sat; and Pope's bill, somewhat amended, was taken up in the Senate in March, 1811, further amended, and finally postponed.

The strength of the forces that defeated these bills boded ill for the future of federal support to internal improvements. True, there was a temporary factor: the Embargo Act was passed while Gallatin was working on his *Report,* and from that time the danger of war was bound to stand in the way of a national improvement effort. But the power of those forces that were to prevent national transportation planning in the postwar years as well was already evident in the debate on the Pope and Porter bills, when states' righters, who opposed federal aid on constitutional grounds, were joined by the representatives of those areas that would receive little or no benefits from the bills. Since geographical discrimination was inevitable in any national effort that deserved to be called a plan, the prospects were dim.[47] Indeed the prospects of federal aid appeared so doubtful by this time that in May, 1811, Gouverneur Morris wrote to Switzerland to inquire whether a loan of $5 million might be obtained on the credit of the state.[48] This was only two months after the 1811 commissioners' report that had taken for granted massive federal aid.

But all hope was not yet abandoned, and in December, 1811, the New Yorkers tried again. De Witt Clinton and Gouverneur Morris visited President Madison as representatives of the New York commissioners and found him enthusiastic about canals. Although he had certain constitutional scruples, they evidently managed to win his complete support, for his message to Congress two days later submitted the New York Legislature's petition for aid, warned that a general system of internal transportation was of great importance to the security of the country at that time, and called for support of New York's canal as a part of that system. The next day the Secretary of the Treasury informed the New York commissioners that, while under current circumstances pecuniary aid could not be given, sufficient grants of land might be made. Later, when the treasury was in better condition, those could be redeemed in cash.

But the New Yorkers' experience with Congress was far less happy. They quickly realized, according to their own later report, that New York's petition could not stand alone, that unless something was proposed for many states at once, the consent of the House was out of the question. On January 10, 1812, therefore, they submitted to the House committee considering New York's petition a bill for seventeen canal projects in thirteen of the seventeen states and in one territory. As the states completed the projects, specified amounts of federal lands in the Michigan and northern Indiana territories were to be turned over to them. Of the 9,900,000 acres to be made available in this way to the states, New York was to receive almost half— 4,000,000 acres for the Erie canal, 400,000 acres for the Champlain canal, and 100,000 acres for a canal in the Illinois territory that would extend from Lake Michigan to the Illinois River. This was a "pork barrel," but it carried a large and clear New York stamp.

The method did not work. The House committee at first seemed to favor the bill, then rejected it on the ground that the resources of the country might be required to support a war. The New Yorkers were sceptical: the reason did not seem to apply to a conditional grant of land which would not take effect until the projects were completed many years later. They discerned other more likely reasons for the failure. Some of the opponents of the bill, so they charged, were advocates of a national bank which hoped to obtain an amendment to the Constitution that would authorize the federal government to incorporate banks as well as make roads and canals without the consent of the states—a highly unlikely development. Others "speak slightly of the canal as a project too vast" or questioned the ability of the state to bear the expense. Still others, though they admitted the vast resources of the state, questioned the ability of the legislature to apply such resources. "These men console themselves with a hope that the envied state of New York will continue a supplicant for the favor and a dependant on the generosity of the Union, instead of making a manly and dignified appeal to her own power. It remains to be proved," so the commissioners defiantly concluded their report on the Washington experience, "whether they judge justly who judge so meanly of our councils." [49]

In their report to the Legislature of March, 1812, the commissioners commented bitterly on these events. They complained of the political necessity of combining the great Erie and Hudson canal project with projects "subservient to local interest." They wrote of the modesty of their request—not an advance, not even an eventual appropriation of money, but merely a grant of land on completion of the canal at the expense of the state. And pessimistic about Congress's future behavior, they concluded that the grapes were sour; that, if the bill had passed, Congress would have obtained a most valuable national benefit in exchange for "a tract of unsaleable land."

Nor could much be expected from the states and territories to be affected by the canal. The commissioners' requests for their financial aid and congressional influence were combined with dire threats of high transit duties if the project were to be accomplished without their help. But of seven replies received, none would promise either grants or loans, only three—Tennessee, Ohio, and Massachusetts—agreed to recommend to their congressmen positive action on the canal's behalf, and one— the territory of Michigan—not only refused all aid but strongly favored the Ontario over the Erie route. Clearly New York could rely only on its own resources.[50]

New York on Her Own

This commissioners' report of March, 1812, marked a new stage in the history of the canal project. Up to that year, the expectation of federal aid had played the decisive role. It had been influential in transforming a visionary idea in the minds of the obscure few into a technically practical project supported by an influential state-wide leadership. It had then enabled that leadership to proceed with all the surveys and cost estimates preliminary to construction without the need of persuading the Legislature and the voters that this enormously expensive, unprecedentedly long canal was worth the sacrifice and the risk it entailed. And it had allowed that leadership to choose the route that promised the greatest long-range benefits despite its far greater length and seeming difficulty. That immunity from the necessity of counting all the costs in the choice was now gone. True, the promoters of the project were to proclaim once again their confident expectation of a federal subsidy. But they knew how uncertain the prospect was, and they had to prepare the public to proceed independently of outside grants or face the possibility of indefinite postponement.

The result was a striking change in the commissioners' approach to the potential revenues of the canal. Before 1812 there had been vague references to the huge potential traffic—$100 million a year of freight "at no distant period" was conceivable, according to the commissioners' 1811 report—but there was no statement that the tolls would cover both capital cost and maintenance and still yield a profit. Even the threats of high transit duties directed at the other affected states and territories in the hope of obtaining aid had not been accompanied by any claims of high profits for the canal. True, an optimistic deduction regarding profits might have been drawn from the rather vague assumptions of rapid settlement of the northwest and from the contention that carrying costs on the canal would be far cheaper than on any alternative transport route. But the fact is that this deduction had not been drawn before 1812. On the contrary, the commissioners' remarks had been extremely sober; they had made no unnecessary assumptions; and it had been enough for them to point to the immense value of the project to the farmers and businessmen along its route and on the Lakes to justify the project.[51]

The 1812 report, on the other hand, was extreme in argument and inspired in tone. As in so much of the country's canal literature during the succeeding two decades, the extreme statements and inspired expectations were justified on the grounds of the remarkable recent development of the country. Any careful projection of recent trends, the commissioners observed, "outruns fancy. Things which twenty years ago a man would have been laughed at for believing, we now see. At that time the most ardent mind, proceeding on established facts by the unerring rules of arithmetic, was obliged to drop the pen at results which imagination could not embrace."

The commissioners then proceeded to outrun fancy. They predicted a rapid settlement of the entire hinterland of the canal that would result in exports from the area equal in volume and value to the current exports of the Atlantic states. It was extremely conservative, then, to estimate that in twenty years time the canal would carry 250,000 tons per year each way. Assuming a toll of $2.50 per ton, the revenue would be $1,250,000. But the interest on $10,000,000 at six percent would be only $600,000. Therefore, even assuming this exaggerated cost, the revenues would not only pay the interest but would soon repay the principal and leave a large profit for the state. With high profits assured, the commissioners could offer an incentive to "those of our citizens who have no immediate interest in the work." For their future taxes would be correspondingly diminished. It is striking that this direct financial profit was so dwelt upon in this 1812 report in which it was first raised, while the long-range, indirect benefits to be acquired from holding a "key to the commerce of the western world" were disposed of in a single sentence.

The need for high profit expectations was probably enhanced by changes in the engineering plan that forced the commissioners to raise their cost estimate for the project. While they quoted extensively from a letter from William Weston which unconditionally endorsed the inclined-plane proposal, they had to modify considerably their 1811 plan, for they found that the inclined plane from Lake Erie could be carried, not all the way to Rome, as they had thought, but only to the Seneca Lake outlet. The canal would then fall by locks to a lower level, proceed on this level until it had to be brought up by locks to the Rome summit, proceed on a level again, then be brought down by a second inclined plane to a basin near the Hudson. These changes, together with some others in standard expense estimates, brought the new cost estimate to $6 million.

Could the state afford such an expenditure? The commissioners laughed (perhaps uneasily) at the very question: "It is almost a contradiction in terms to suppose that an expenditure of five or six million, in ten or a dozen years, can be a serious consideration to a million men, enjoying one of the richest soils and finest climates under heaven." Add to this the fact that the population doubles every twenty years, and the ease of financing the project by taxation becomes "one of those evident propositions which argument may rather obscure than elucidate." For the burden would amount to a mere $500,000 a year—$2.50 per family—while the wealth of the state, according to the commissioners, amounted to about $500 millions and was growing rapidly. But even this burden could be reduced by using the credit of the state. The commissioners had ascertained that despite the scarcity of money in wartime Europe a loan of $5 million could be obtained there for a term of ten or fifteen years at six percent interest.

The commissioners argued emphatically against a postponement of the project. Will time render the matter any easier? they asked. "Will it alter the shape of the country? Will the land to be used for the canal cost less when it shall be planted . . . ? Will timber and lime be cheaper when wood, now worth nothing, shall have grown dear?" The point was well taken; there was a great cost advantage in building a canal through unsettled areas. The commissioners therefore pleaded for immediate legislative action.[52]

Their plea was successful. In June, 1812, an act, that had been passed in each house by a small majority, authorized the commissioners to borrow up to $5 million on the credit of the state. But this was not a decision to construct the canal. Whereas there existed a favorable opportunity to procure the loan, the act states, and whereas it was desirable that the state possess the means of opening an inland navigation "if upon full examination by a competent and practical engineer, and mature deliberation, the Legislature shall hereafter deem it expedient to undertake that interesting work," therefore the commissioners were authorized to procure the loan and invest it until it might be needed. They were also authorized to negotiate purchase of the rights of the Western Company, but the purchase was to take effect only after an experienced engineer examined the route of the proposed canal and pronounced it practicable, and the Legislature authorized the commissioners to begin construction.[53]

The commissioners seemed confident that the project would be authorized. They quickly wrote to England for an engineer, hoping he would arrive by the fall, and they proceeded to apply to the proprietors of large tracts in western New York for grants of land. As for the loan, they were certain that they could procure it at any time. In July, Thomas Eddy wrote to Joseph Ellicott, the agent of the Holland Land Company, that much depended on the size of the grant by this company, for other large proprietors would follow its example. "If *the whole* should amount to something considerable, there is no doubt that the Legislature would agree to commence the work." [54]

Curiously, all this optimistic activity occurred in the midst of war. The debate in the Legislature on the bill authorizing the $5 million loan had coincided with the Congressional debate on President Madison's June 1 war message, and the bill had passed on June 19, the very day that Madison proclaimed a state of war. The commissioners proceeded on the assumption that the war would not affect their project; indeed, there was evidently some belief that, in view of the military value of the project, the war would increase the interest of the federal government in its completion.[55]

But these were serious miscalculations. Because of the war, the British engineer could not come over, the attempt to obtain a European loan failed, and, in the summer of 1813, "military operations which are not favourable to internal improvement" forced the commissioners to suspend their surveys.[56] Events in the Legislature were even more discouraging, for they implied more than a mere temporary suspension of activities. In February, 1813, a resolution directing the commissioners to report to the Legislature on their progress, particularly in obtaining a European loan, was rejected in the Senate by a vote of twelve to nine. Even Senator Platt voted against the resolution. A year later, the Assembly rejected a motion to suspend the loan authorization of the 1812 act until one year after the termination of the war; instead, the Legislature repealed the provisions outright.[57]

This wave of opposition to the project continued into the postwar period. Most important was the general recognition that the state would have to bear the entire burden, for this was bound to increase sectional opposition to the canal. At the same time, the military expenditures and the postwar economic difficulties seemed to put the cost of such a project far beyond the state's power. To many it seemed that the shorter and cheaper Ontario canal—or even mere improvement of the watercourses—was the natural thing for a state of scarcely a million of population so recently involved in war.[58] This feeling was reinforced by the technical difficulties of the inclined-plane technique. The commissioners had considerably modified the idea in their 1812 report, and in their 1814 report retreated further by leaving the issue for future investigations to decide.[59] But the effects of the ridicule aroused could not be eliminated so easily. The projectors of a plan so radical as that of the Erie needed above all a reputation for practicality, but that reputation was in good part destroyed by the national reception given the inclined-plane proposal.[60] For all these reasons, the reaction against the canal idea was extremely strong, so strong indeed that many of the project's supporters abandoned all hope of its revival.[61]

The Project Carried Through

Yet, only two and one-half years after the war ended, the state began construction of the Erie Canal. The popular movement that was to carry the project to success initiated the first realistic state-wide public discussion of the canal issue. No longer was the cost problem hidden behind the federal purse strings; no longer did the canal's supporters call for investigation rather than decision; and no longer was the debate primarily in the Legislature. Instead, De Witt Clinton took the issue to monster mass meetings, where he demanded immediate construction on the Erie route and declared openly that the state must be prepared to finance it all.

Clinton's role was crucial. Not the type to be attracted to visionary proposals, he had displayed no interest in the canal question before 1810. Even after his support had been obtained and he had been made a commissioner, he did not take the leading role.[62] But in 1815 two troughs coincided: the canal movement was in its most depressed state and his political career seemed at an end. By taking the public leadership of the canal forces, he saved both the project and his career.

Clinton's faction in the Democratic party had been smashed by the Tammany organization. Aided by the lukewarm attitude of the New York Federalists towards the war, Tammany took control of the state and in 1815 forced Governor Daniel Tompkins to remove Clinton as mayor of New York City. Clinton was left without office, without a party or the fraction of one, and deeply in debt. His political aspirations seemed doomed.[63]

Clinton's feud with Tammany dated back to 1802. Burdened by a cold and haughty personality, he was incapable of machine politics, despised Tammany's careful construction of a ward-by-ward organization, and depended instead on the force of his ideas, daring maneuvers and alliances, and his personal popularity, particularly among the Irish.[64] For this defeated lone wolf, the cause of the Erie Canal was an ideal lever for the leap back to political power. As an intellectually gifted man, he saw clearly the drama and potentialities of the project; and as a personal—not a machine—politician, he could forge a movement on a single issue that drew supporters from all the parties and factions of the state. It is quite possible that without the canal issue, Clinton's political life would have been finished; it is also possible that without Clinton, the canal would have been postponed for many years—perhaps until it was superseded by a railroad project. Together, however, they won the state with remarkable rapidity.

Clinton began under discouraging circumstances, for the general pessimism regarding the prospects of the canal tended to inhibit all action. In the fall of 1815, however, Clinton, Jonas Platt, and Thomas Eddy determined to make one more effort. Invitations were sent to about one hundred prominent men of New York City, and on December 3 a meeting was held under the chairmanship of William Bayard. In his address, Platt urged the formal abandonment of the inclined-plane method. Clinton was appointed chairman of a committee to memorialize the Legislature and took full advantage of the opportunity. His "New York Memorial" was a model of lucidity, logical power, and thorough factual knowledge. The copies sent throughout the state evoked a response of remarkable power in view of the pessimism that had just previously prevailed. Public meetings in almost every city and village along the route of the proposed canal produced similar memorials. Over one hundred thousand signatures were sent to the next legislature. Wealthy Federalists —large landholders in the west and great merchants in the cities

—as well as western farmers were attracted to Clinton's banner. The Irish, whose numbers had recently increased greatly, rallied to Clinton because of past favors and in reaction to Tammany's nativist policy (which was not reversed until 1827). Large merchants and western farmers, aristocratic landholders and poverty-stricken immigrants—all united behind Clinton and his great canal project.[65]

Clinton's "Memorial" was a classic statement of New York's transportation opportunities and needs, a culmination of the series of such statements that began with Cadwallader Colden's memoir of 1724. Canals, he declared, by cheapening the cost of transportation have the effects of labor-saving machines; by diminishing distances they encourage the cultivation of remote areas; by increasing internal trade they destroy local monopolies; by augmenting individual wealth, encouraging the growth of population and the construction of towns, and extending foreign commerce increase the market for all goods. Hence the prosperity of a country was proportional to the extent of its inland navigation. The United States was particularly fortunate in this respect, for the size of its inland lakes exceeds that of many of the great seas of the Old World. To unite these "Mediterranean seas" with the ocean was therefore of the first importance. Nature had already done part of the work: there were many streams extending west from the ocean. But only the Hudson broke through the Blue Ridge and then found no mountains interposed between it and the Great Lakes.

Such a geographical situation, Clinton maintained, had staggering possibilities. "If we were to suppose all the rivers and canals in England and Wales, combined into one, discharged into the ocean at a great city, after passing through the heart of that country, then we can form a distinct idea of the importance of the projected canal." New York City would emerge with so great an advantage over its most powerful rivals for the western trade that it could easily become "the greatest commercial city of the world."

The "Memorial" opposed the Ontario route with the familiar argument concerning Canadian competition. But this was not made the decisive argument; even if the Ontario route also were to bring the commerce of the west to New York City, the Erie route should be preferred. An "invincible argument" was that the Erie would "diffuse the blessings of internal navigation" over a far greater area of the state. In addition there were the advantages of a single canal over canal-lake-canal system: freedom from the winds and waves of a lake and freedom from transshipments. This last advantage was strongly emphasized. Since lake vessels could not use the canals, the Ontario route would require two more transshipments than the Erie, making a total of five transfers of goods between Lake Erie and the wharves of New York City. This meant longer delay, higher labor costs, more damage to goods and susceptibility to theft, and greater capital requirements in terms of boats, docks, and storage facilities. A final argument dealt with the great lockage within a short distance required by a Niagara canal—forty-five locks in a distance of 450 feet—and the difficulties of the Ontario canal between Oswego and Oneida Lake. In contrast there was the ease of construction of the Erie. Only sixty-two locks in all would be required; every summit level had plenty of water; the rivers could easily be crossed by means of aqueducts; the one area that rose higher than Lake Erie could be handled by deep cutting. In sum, there were no major difficulties. As for the unprecedented length that bothered so many people, "If a canal can be made for fifty miles, it can be made for three hundred."

On the basis of European and American canal experiences, the "Memorial" put the cost of the Erie at $6,000,000. Gallatin had estimated the cost of a canal to Lake Ontario at $2,200,000— Clinton cited this estimate—and he had put the cost of a canal around Niagara at $1,000,000. In calling for the Erie route, then, Clinton was (implicitly) advocating almost a doubling of construction costs. Indeed, Clinton called for immediate construction even if the cost of the Erie were twice as large as his estimate.

The work might be accomplished either by the state or a private company, but, if the latter method were used, the "Memorial" warned, public subsidy would still be required and there would be danger of high tolls that might well ruin the project. State construction would impose a very small burden; since the project would take ten or fifteen years to complete, only the interest cost on borrowings of a half million dollars per year would be required. Furthermore, lands donated by individuals, which already amounted to 106,632 acres, would undoubtedly exceed a million dollars in value. In all likelihood, the augmented revenue of the public salt works and the increased price of the state lands would more than cover the interest charges, while light transit duties on completed sections of the canal would easily extinguish the debt. There was no reference to the possibility of a federal subsidy.

The "Memorial" ended with a passionate plea for immediate action. The time was ideal, for the expenses of transportation during the recent war had demonstrated the vital need for the work, the canal was the "sovereign remedy" for the current invasion of foreign goods, and the men dismissed from the army and thrown out of work by the shocks to foreign commerce were available for the work of construction. Delay, "the refuge of weak minds," would be disastrous, for the value of the lands through which the canal would pass increased day by day, and the interests opposed to the project were increasing as substitutes, such as turnpikes and short canals, were adopted and villages were laid out in accordance with these expedients. Most important, delay threatened the national unity; only New York, a state "both atlantic and western" had the power to prevent the calamity of a split between east and west.

Nevertheless, delay was a natural response of the Legislature. The wave of canal enthusiasm engendered by the New York "Memorial" met with a storm of opposition based upon sectional interests, political hatred of Clinton, and, not least, a skepticism of so large a project. It is noteworthy that the skepticism was shared by Paul Busti, the American agent of the Holland Land Company. He recognized that the Dutch investors he represented would benefit enormously from such a canal; and he gave to the state a large—but carefully chosen—sector of the company's vast holdings in western New York. But he considered the donation a mere public relations gesture. "The magnitude of the undertaking of the canal is so great that it is impossible for me to believe that the work will ever be perfected," he wrote to his Batavia agent in October, 1816.[66]

These skeptical reactions were strengthened by the fact that the rivals of New York to the south were placing their hopes in the tried and true method of road building. The federal government's Cumberland Road was advancing westward; it reached the Ohio at Wheeling in 1818. Pennsylvania had embarked upon a great road-building program in 1811 when the state appropriated $825,000 to aid the turnpike companies. The stone-surfaced Pittsburgh Pike between Philadelphia and Pittsburgh, completed in 1820, was to become one of the major carriers of freight to the Ohio Valley.[67] Everyone knew what these roads would do; few were certain of the effects of a long-range canal. It is significant that the Erie Canal project produced no anxiety—and much ridicule—among the rivals of New York City until the immense receipts of the partially opened canal had forcibly demonstrated its success to all.[68]

The sectional opposition had a number of focuses. The Long Island counties and those on both sides of the Hudson River anticipated the competition of a flood of cheap western agricultural products in the New York City market. The canal's supporters, unable to deny this in view of their predictions of the volume of agricultural traffic on their project, replied that the canal would induce so rapid a growth of the city that the demand would at least keep up with the greater supply.[69] The southern tier of counties produced a second center of opposition. These counties along the Pennsylvania border complained that the canal would be of no use to them, that they would still have to float their produce at great hazard and expense down the Pennsylvania rivers. To appease them, Clinton referred to future branch canals from the Erie and promised to work for a road through the southern tier from the Hudson to Lake Erie. Finally, the counties around the eastern end of Lake Ontario, which had been forced to do their trading with Montreal, demanded the Ontario route with its outlet at Oswego.[70] Of course, the pairing of the Champlain with the Erie project reduced somewhat the size of the sectional opposition.

When the legislative session opened in February, 1816, it received the multitude of canal petitions with their one hundred thousand or more signatures. But it also received a message from Governor Tompkins that contained only a tepid reference to a canal with an unspecified route. Clinton suspected that the governor and his other Democratic enemies were preparing a strong bid for the Ontario route as a method of destroying the project associated with his name.[71]

The commissioners' 1816 report to the Legislature hammered away at the disadvantages of such a course. Only the Erie route could save the infant commerce of western New York from a commercial connection with Montreal. They suggested that construction begin with the Rome-to-Seneca River section, for it would immediately divert the trade passing down the Oswego River to Lake Ontario and Montreal, and it would yield the most revenue relative to cost. Now that the war was over, they could assure the legislature that a loan could easily be obtained. There was no mention of the inclined plane.

There was, however, the problem of finding an engineer, and for the first time the commissioners suggested the feasibility of hiring an American. Their negotiations with William Weston had fallen through—he would not leave England even for a proffered salary of $7,000 a year—and there were few such capable men in Europe. While Gouverneur Morris insisted on the necessity of an experienced engineer—which meant a European—Joseph Ellicott believed that foreigners should be avoided: "They know very little about the management and conducting of business economically in this country and the truth is the laying out of a path for a canal neither requires conjurers nor wizzards; practical nature is everything." Whether to avoid a long delay or out of preference, the commissioners decided to put their confidence in their own surveyors, who were certain that they could locate and construct the canal. The decision met with severe criticism. "On the Assembly floor it was tauntingly asked, 'Who is this James Geddes and who is this Benjamin Wright . . . what canals have they ever constructed?' " This issue of technical competence was a central one partly because of the canal's unprecedented length. Many contended that the spirit level would be inadequate; that even an experienced engineer could not lay out such long levels without serious error.[72]

A bill authorizing a loan of $2 million for immediate construction of the Champlain Canal and of the Erie between Rome and the Seneca River (the "middle section") finally passed the Assembly when the lands adjacent to that middle section were made liable to taxation for the building up of the sinking fund—a concession to the sectional opposition to the canal. But the Senate, on Martin Van Buren's motion, deleted the authorization of construction on the grounds that more detailed and accurate surveys were required before a decision could be made. The Assembly was forced to accept the amendment; as passed, the bill merely authorized the commissioners to conduct further surveys, apply for donations of lands and for aid from other governments, and ascertain whether a loan could be procured. The new commissioners did not include Gouverneur Morris because of his sharp disagreements; he died a few months later.[73]

The commissioners, with Clinton as president, spent the year gathering detailed survey information in an attempt to provide the answers to every possible doubt and technical objection that might arise in the next meeting of the Legislature. They even sent a delegation to examine what they considered to be the best artificial navigation in the country, the Middlesex Canal. Their elaborate report, presented to the Legislature in February, 1817, called for a 353-mile canal with 77 locks and a total rise and fall of 661 feet. The cost was estimated at somewhat less than $5 million—about $13,800 per mile. They had not yet ascertained whether a loan could be obtained in Europe but had begun the negotiations for one. Although their efforts to obtain aid from Congress and the states had as yet led to nothing, they assured the Legislature that the power to levy high transit duties on the completed canal would eventually secure a greater fund than any they could expect on a voluntary basis. But they still hoped for cooperation. On this basis, the debate began in the Legislature.[74]

In the following month Jacob R. van Rensselaer, the great landholder of Claverack, threw a bombshell into the proceedings by offering to form a private company to build the canal. The company would deposit $1,000,000 as a guarantee that the entire canal would be completed on the exact plan of the commissioners (with one minor exception). The state was offered three alternative forms of payment to the company: (a) $10,000,000; (b) $7,500,000 plus the tolls for twenty years, the tolls not to exceed two cents per ton-mile; (c) $5,000,000, the tolls for twenty years (at the two-cent maximum), and half the tolls forever thereafter. If the Legislature agreed to a toll of $5 per ton over the route for the period after the first twenty years, the company would pay the state annually three percent of the $5,000,000 or would simply return $2,500,000 plus interest. From the viewpoint of hindsight the offer was a conservative one; if it had been accepted, particularly in its second or third variant, Van Rensselaer would have made a fortune. But at the time the offer was important as a demonstration of confidence in the profitability of the project. The Legislature's joint committee on canals reported the offer and rejected it: the state must retain complete control of both construction and operation.[75]

This joint committee had delayed its report to the Legislature until March 19 because of the appearance of a last hope of federal aid. A bill creating a fund for internal improvements from the bonus paid the government by the Bank of the United States and from all future dividends from the government's bank stock passed both houses of Congress. If distributed in accordance with population, New York would have received $85,000 to $90,000 per year. This would have been small relative to the expected annual expenses of construction, but, in the view of the canal leaders, it would have guaranteed passage of the Erie Canal bill. On March 3, however, Madison vetoed the "Bonus Bill" on constitutional grounds.[76]

New York, then, would have to pay for it all. But who in

New York should pay? The bitter sectional struggle in the Legislature centered on this issue. It was proposed that the required loans would be backed by a "Canal Fund," which would accumulate from the donations of land and any other grants, from tolls, and from taxes the funds for construction and for payment of interest and principal on the loans.[77] The taxes, said the representatives of New York City and the western district, should be state-wide in their incidence, for those who used the canal would be taxed when they paid the toll. But the majority saw it differently. Both the bill that passed the Assembly in 1816 and the bill proposed in 1817 placed a large proportion of the tax burden on the city and the west. They called for diversion to the Fund of much of the city's auction duties, thereby reducing the funds for poor relief that the city obtained from this source; and they imposed a tax on the salt produced by the western district, a tax on real and personal estates in New York, Albany, and the towns and counties along the line of the canal, and a passenger tax on steamboats. In addition, part of the state's public lands were to be diverted to the Fund. But New York and the west seemed unwilling at first to bear any special taxes at all. As an infuriated Assemblyman put it,

Will not all the productions of this vast and fertile territory, go to New-York? . . . If this canal is to be a shower of gold, it will fall upon New-York—if a river of gold it will run into her lap. . . . Are we to give her the means of enriching herself, beyond all former example, and of monopolizing the trade of the whole world, and she pay nothing in return? . . . Is then our direct tax to be doubled? [78]

The reasoning seemed justified, though with every New York City representative eventually voting against the canal and its shower of gold, reasoning may have been irrelevant. In the end, the principle of special taxes was maintained, but there were a few minor changes. A tax totaling $250,000 on lands within twenty-five miles of each side of the canal was substituted for the tax on towns and counties but was never collected; the provision for the granting of state lands was deleted; the proceeds of future state lotteries were added; and an amendment proposed by Van Buren in the Senate permitted loans on the credit of the state rather than on the credit of the Canal Fund. The 1817 bill, as passed, provided for a "Board of Commissioners of the Canals Fund" which was allowed to issue transferable securities payable at times to be determined by the board for loans that, together with the income of the Fund, were not to exceed $400,000 per year.[79]

Why did the bill pass in 1817 when it had failed in the previous year? The stumbling block in 1816 had been the Senate, and in the Senate it had been the power of the state Democratic organization led by Martin Van Buren. With Clinton's career so closely tied to the success of the canal bill, the opposition of the organization Democrats seemed inevitable. Nevertheless, in 1817 Van Buren suddenly announced that his previous objection to canal construction—the lack of detailed information—had been overcome by the last report of the commissioners. This move was probably decisive: the vote in favor of the bill was eighteen to nine, and five of those eighteen were zealous anti-Clintonians. Judging from the report of his speech in the Senate, Van Buren shifted not only because he was actually convinced that the canal should be built but also because the project had become too popular to oppose. As he put it, "Our tables have groaned with the petitions of the people. . . . Twelve thousand men of wealth and respectability in the city of New-York last year petitioned for the canal." The bill had twice passed the Assembly, "the immediate representatives of the people"; hence, "we are bound to consider that the people have given their assent." If the bill did not pass this time, however, the project would necessarily be postponed many years. This recent opponent even revealed that he was a canal enthusiast. He considered this vote the most important in his life because the canal "would raise the state to the highest possible pitch of fame and grandeur." [80]

That the entire New York City delegation nevertheless held out against the canal in both Assembly and Senate has been considered merely as evidence of the intensity of partisan passions in the city's politics. It is true that those votes were directed at Clinton and his ambitions; they certainly did not represent any important mercantile opinion in the city against the project.

Of all the daily newspapers there, only the extreme Tammany organ, the *National Advocate*, opposed the canal during 1817. But mercantile opinion was not necessarily the only important opinion. An historian of the Tammany Society distinguishes three groups in its membership during this period: the early leaders whose recently acquired wealth "could not quite secure them admittance to that stiff aristocracy above them"; the much larger body of small tradesmen; finally, the mechanics and laborers.[81] These were the classes that found attractive the embittered denunciations of the aristocracy, the self-righteous indignation at lower-class vice, and the prejudiced nativism that were the hallmark of the *National Advocate*. Utterly foreign to the rank and file of this organization were the broad view, the long-range vision, and the large expenditures involved in a great transportation project; and utterly hateful to them were the extremes of the Clintonian alliance—aristocratic landowners on the one hand, Irish immigrants on the other. It was their representatives who tried to prevent that "river of gold" from flowing into their city's lap.[82]

After its passage, however, there was still the possibility of a veto that could kill the bill, for the two-thirds majority in both houses necessary to override was probably out of the reach of the canal party. The veto power was at that time lodged in the Council of Revision, consisting of the (Acting) Governor, the Chancellor, and the three judges of the Supreme Court. The Council's discussion revealed a three-to-two split against the measure; the bill was on the verge of destruction. But toward the close of the discussion, the former Democratic governor, Vice-President Tompkins, walked in and informally joined the debate. Tompkins was against the bill and gave as one of his reasons the prospect of a renewal of the war with Great Britain—"England will never forgive us, for our victories." The credit and resources of the state, he declared, should be used, not in this chimerical project, but for the building up of the arsenals, the arming of the militia, and the erection of fortifications. No other argument could have so aroused the hostility of the Federalist Chancellor, James Kent. Up to that moment he had opposed the canal bill principally on the ground of its vast expense, but he was now convinced that thrift was no longer an alternative. According to the later account of one of the members of the Council, Kent declared in great agitation that "if we must have war, or have a canal, I am in favour of the canal," and the bill passed.[83]

Thus, to the very end, the War of 1812 and its consequences powerfully influenced the fate of the canal project. The war itself forced a postponement. But the failure of the attack upon Canada reinforced the arguments against the Ontario route, while the great difficulties of military operations on New York's frontier caused by poor transportation facilities gave an impetus to the national movement for internal improvements. There is some justification, therefore, for one historian's conclusion that the canal was one of the results of the war. It is conceivable that if the convention of 1818, which settled the international boundary question, had been held in 1816, the canal would have been postponed for some time.[84]

A postponement might have been permanent. Jonas Platt thought it probable that, if the measure had been defeated again in 1817, it would never have passed the Legislature. At no future period, he pointed out, could the work have been accomplished as cheaply and easily. With the west rapidly filling up, land and water values were soaring, rival routes were multiplying, local interests were daily more powerful and clamorous.[85] Furthermore, even if it were not permanent, any long postponement would have had a profound effect upon the development of New York and its principal city. For the railroad, just appearing on the horizon in 1817, was soon to conquer the mountains and thereby eliminate a good part of that fateful New York advantage over its rivals on the eastern seaboard.[86] As it was, however, the timing was perfect. At the very beginning of a half century of almost uninterrupted peace and economic expansion, New York City not only seized the initiative in drawing British goods to her port by means of auction sales and an innovating packet service, not only did she throw a net of shipping lines up and down the coast, but she also, by means of the canal, tied to her wharves the products of the most fertile continental area on earth at a time when no other city could possibly do so.[87]

Judging from the history of the canal project, this combination of events was far from accidental. True, Chancellor Kent's reversal in the Council of Revision was probably the result of an incidental factor. But the forces that brought the project to that point must have been of a far different character. It must be remembered that, with the single exception of the Languedoc Canal, built by the imperial government of France in the seventeenth century, no one in the western world had ever before built a lock canal one-tenth the size of the Erie; and that most of the European canals had been dug through regions fully settled and cleared. The remarkable fact is that a population of little more than a million took on an expected burden of $5–6 million in order to build an immensely long canal through a sparsely settled wilderness without the aid of a single experienced engineer. In the process of construction, the country's first hydraulic cement was produced, new types of excavating and tree-handling machines were invented, and a group of engineers was trained that was to direct the building of much of the country's transportation network in succeeding decades.[88]

A project of such boldness cannot be explained merely by reference to qualities of foresight and statesmanship. As mentioned above, Thomas Jefferson reacted to news of the canal proposal in 1809 with the remark that it was a century too soon, that "it is little short of madness to think of it at this day." Thirteen years later, when De Witt Clinton reminded him of it, Jefferson admitted the error, adding that "many, I dare say, still think with me that New York has anticipated by a full century, the ordinary progress of improvement." But his inquiring mind went on to raise the most interesting problem of all regarding the Erie Canal:

> This great work suggests a question, both curious and difficult, as to the comparative capability of nations to execute great enterprises. It is not from greater surplus of produce, after supplying their own wants, for in this New-York is not beyond some other states; is it from other sources of industry additional to her produce? This may be—

or is it a moral superiority? a sounder calculating mind, as to the most profitable employment of surplus, by improvement of capital, instead of useless consumption? I should lean to this latter hypothesis, were I disposed to puzzle myself with such investigations; but at the age of eighty, it would be an idle labour, which I leave to the generation which is to see and feel its effects.[89]

That question of comparative capability, still unanswered, will confront us again in our next example, that of the Pennsylvania Mainline.

II. An Imitative Public Improvement: The Pennsylvania Mainline

BY JULIUS RUBIN

Pennsylvania's drive for a modern transportation line to the west began with the success of the Erie Canal. True there had been earlier proposals—the possibility of a canal between the Delaware and the Ohio had been discussed as early as 1786—but up to the 1820s New York's seaboard rivals had confidently relied upon their turnpikes to maintain their position in the western trade.[1] That policy had been successful, for the shortest route to the west led across Pennsylvania, Maryland, and Virginia. Pittsburgh, Cincinnati, and Wheeling—the points of transshipment on these routes—became important commercial centers at a time when Buffalo, Cleveland, and Detroit were mere frontier settlements. In this respect, New York City was at a disadvantage during the turnpike era.[2]

Of all these rival areas, Pennsylvania had the most reason to concentrate chiefly upon land transport. Whereas New York had the Hudson and Mohawk, and Maryland and Virginia had Chesapeake Bay and the Potomac, Pennsylvania, like New England, was poorly equipped with rivers suitable to transportation to the west. The shallow Schuylkill and Susquehanna rivers were constantly interrupted by rapids, boulders, bars, and islands; and, though the Susquehanna drained half the territory of the state, its course tended to bring the traffic to Baltimore.[3]

In the building of the turnpikes, Pennsylvania had excelled.

In 1811 the state appropriated $825,000 to the aid of turnpike companies and by the end of 1821 these companies had completed 1,807 miles. Pennsylvania was then equipped with excellent trunk lines from Philadelphia to Pittsburgh and to various points on the New York State border, and from Pittsburgh to Lake Erie. In addition, the National Road passed through the southwestern corner of the state on its way to Wheeling and a branch connected it to Pittsburgh.[4]

The two principal east-west turnpikes gave to Philadelphia and Baltimore a favored position in the western trade. It is estimated that in the early 1820s the Pittsburgh Pike carried about 30,000 tons of goods annually, the National Road about 10,000 tons. But these were overwhelmingly manufactured goods shipped westward. The bulky agricultural produce of the west continued to use the Mississippi route to New Orleans. Furthermore, the growing use of the steamboat for upriver carriage on the Mississippi and Ohio rivers even threatened to deprive the turnpikes of their westward trade. New Orleans was still a dangerous rival of the eastern seaboard cities.[5]

This situation was transformed by the construction of the Erie Canal. Here was a transportation method far superior both to the turnpike and to the river, with or without steamboats. The canal immediately took over from the turnpikes a part of the westward trade and, when the immigrants it transported had built up the northern midwest, it carried their agricultural produce back to the east. For the first time, west and east were linked by a direct two-way trade.[6] These dramatic effects, signaled as early as 1823 by the remarkably high earnings of the partially completed canal, sparked a national canal mania. All the major seaboard rivals of New York initiated extensive surveys for competing canals to the west.

In this field, however, New York had an immense advantage. A 3,000-foot mountain barrier stood in the way of all the rival cities and parted only for New York. So great were the technical difficulties and economic disadvantages of building a canal over a mountain range that no canal was ever completed over the Appalachians. By the 1830's it was clear to many that only by a return to land transport and the utilization of the new technique of the graded track combined with the locomotive could the rivals of New York conquer the barrier and gain access to the western trade.[7]

A few Pennsylvanians foresaw that the railroad, primitive and unproven as it was in 1825, would nevertheless be the answer to the problem. But there was little time to find out. If they did not soon get a competing line across the mountains, they felt, the channels of commerce would be set, perhaps permanently, in the direction of Philadelphia's rivals. And this danger appeared at a time when the center of economic interest was shifting from the ocean to the developing interior, from foreign to domestic commerce.[8] A fear of stagnation, even decline, dominated their thoughts in this period.

If it were to be done, then, it would have to be done quickly. But a Pennsylvania decision to build an interregional railroad in 1825 would have been far more daring than the New York decision to build a long-range canal in 1817. The bold stroke that had produced the Erie and its success—the abandonment of the traditional technique of river improvement with short canals and the projection of a full canal independent of the direction of the watercourses of the state—had been prepared by decades of large-scale visions and small-scale experiments. The Pennsylvanians did not have decades; indeed so great was the sense of urgency of the improvement leaders there that they were reluctant to spend even a year upon fact-gathering and discussion. Thus there was a crucial difference between the New York and the Pennsylvania situations. Both states built a line to the west that involved large expenditures, complex engineering problems, and long-range developmental goals. But whereas New York embarked upon her project after long consideration and experimentation, Pennsylvania plunged into hers under extreme duress, bitterly aware of lost marches and of the need to overtake a powerful rival.

This disadvantage was compounded by the far greater power of sectional forces in Pennsylvania. New York's population has always concentrated in a narrow band of settlement. It is estimated that in 1814, sixty to seventy percent of the state's population lived along the borders of the Hudson, the Mohawk, and other waterways and lakes. In addition, the sparseness of settlement in the far western part of the state meant that even intense sectional disagreements there—over the route and terminus of the Erie Canal, for example—never endangered the project. To these factors, and to the aristocratic structure of New York society, have been attributed the state's ability to concentrate its resources early upon two strategic projects, to finance them conservatively, and to build them efficiently. Settlement in Pennsylvania, on the other hand, had not been so long held up and had no such central tendency. As a result, powerful forces opposed the concentration of the state's resources upon a single interregional line. Several counties in eastern Pennsylvania played a role similar to that of the Hudson River counties in the New York canal agitation: Lancaster was content with its turnpike to Philadelphia; Lehigh, Lebanon, Perry, and Union counties felt that the Susquehanna River, the Union Canal between the Susquehanna and the Schuylkill (soon to be completed), and the Schuylkill navigation to Philadelphia would be adequate to their needs. Other sections had developed "foreign" allegiances. Along Pennsylvania's southern border, farmers marketed their produce in Baltimore or in the neighboring counties of Maryland and Virginia; in the northeastern section of the state they traded with New York. It was in reaction to the tendency of these sectional interests to scatter the resources of the state on local projects that the term "mainline" was adopted for Pennsylvania's answer to the Erie Canal.[9]

The directing force behind the movement for this mainline between Philadelphia and Pittsburgh was the Pennsylvania Society for the Promotion of Internal Improvement in the Commonwealth (hereinafter referred to as "the Society" or "the Pennsylvania Society"). Fathered by Mathew Carey, the well-known Philadelphia publisher and propagandist, its stated purposes were, first, to disseminate knowledge throughout the state regarding the transportation problem and the urgent need for improvements, with the purpose of forcing legislative action;

second, to collect information on transportation possessed by other states and foreign countries. An "Acting Committee" chaired by Carey directed the day-to-day work.[10]

As one of its members put it, the Society was organized in response to the sad contrast between New York's great advance and Pennsylvania's indecisiveness and delay. There were many friends of internal improvement in Pennsylvania, "but they had no interchange of opinion, no concerted and arranged plan of measures and operations." That concerted plan was supplied by a small, geographically concentrated group. Formally organized in December, 1824, with forty-eight of the leading citizens of Philadelphia as charter members, it was prevented from increasing in size during 1825 by the substantial initiation fee of $100.

The importance of the Society can best be judged from its activities. By March of 1825 it had published eight technical papers on turnpikes, canals, and railways in editions of one thousand copies each. Its essays were reprinted in newspapers throughout the state. Committees were set up in every county to discuss the improvement issue and to memorialize the legislature. Citizens were circularized over and over again in favor of a modern transportation line between Philadelphia, Pittsburgh, and the Great Lakes. A prominent Philadelphia architect and engineer, William Strickland, was commissioned to investigate recent European improvements, particularly in railroads. Strickland and his assistant, Samuel Kneass, left for England on March 20, 1825; beginning in August, his mailed reports were distributed by the Society in pamphlets and broadsides, and these were reprinted in the press throughout the state. Clearly, the improvement forces of the state did not lack leadership during the discussion of 1825 that produced the Pennsylvania Mainline. They were led by a group of relatively wealthy men with unequaled access to the best technical data available, a group obviously able to give the movement direction and focus.

The Society was organized at a time of national canal agitation. The Erie Canal, partially open, was already demonstrating its tremendous potentialities. Further south, the agitation for a Potomac canal to the Ohio was achieving its first results—by

February, 1823, surveys to Cumberland had been completed, the Potomac Canal Company had been organized, and a Congressional committee had recommended to the attention of the House the impressive potentialities inherent in such a line.[11] As a result, the Pennsylvania Legislature, formerly passive, suddenly acted with great dispatch. A mainline bill that had been pigeonholed since February, 1823, was reported out favorably in December in recognition, as the House committee put it, of the urgent need to keep pace with New York and Baltimore. The bill, signed into law on March 27, 1824, called for the appointment by the governor of a three-member "Board of Canal Commissioners" to explore three possible routes for a Philadelphia to Pittsburgh canal.

The years of delay were now succeeded by months of extreme haste. The Board was appointed within four days of the signing of the bill and was required to report to the governor before the next session of the Legislature in December, giving it very little time to make the required surveys. Unable to obtain a competent engineer, the Board began its work without one.[12] Nevertheless a majority report was issued in February, 1825, signed by two of the three commissioners, Jacob Holgate and James Clarke. It announced the "perfect practicability" of a continuous canal from Philadelphia to Pittsburgh, providing that a four-and-one-half-mile tunnel be driven under the summit of the Alleghenies. The commissioners recommended the route along the Susquehanna and Juniata rivers in the east, along the Conemaugh and Allegheny rivers in the west; they estimated the cost of construction from Middletown on the Susquehanna to Pittsburgh at $3 million (including $480,000 for the summit tunnel); and they thought it could be completed in six years.

The commissioners suggested that the mainline at first use the Union Canal and the Schuylkill Navigation as the link from the Susquehanna to Philadelphia, but stressed that this could be but a temporary expedient, for the canal was circuitous—the country was hilly and short of water—and its narrow locks could never handle "the whole commerce of the west" when it came streaming through the mainline and combined with the river trade at the Susquehanna. A larger, more direct canal would have to be built, but as their surveys in this area were even more sketchy than elsewhere, they did not estimate the cost. They did, however, estimate that it would have required a lockage of 741 feet in about 125 miles. On the western end, an eventual extension to Lake Erie was recommended, for then "we can take the cream off the lake trade before the icy fetters of winter are loosened from the New York canal."

The tunnel would be 754 feet below the summit of the Alleghenies and thus save a large amount of lockage in the most difficult section; nevertheless, 2,595 feet of lockage would remain in the 270 miles between the Susquehanna and Pittsburgh, 1,329 feet of it in the mountain section. This compared with a total of 689 feet of lockage on the Erie Canal. The suggested canal prism was slightly smaller than that of the Erie: 24 feet wide at the bottom, 40 feet wide at the water line, and 4 feet deep, compared to the Erie's 28 by 40 by 4.

The commissioners conceded that New York "had some advantages over us in lockage." But Pennsylvania had other and greater advantages, they thought: convenience of materials along the route, the current low rate of interest, lower wages, cheaper provisions, more experienced workmen, the possibility of avoiding the construction errors made on the Erie, and a warmer climate that would permit a longer season of navigation. Hence they were extremely optimistic. The tolls from this great work, they predicted, would eventually entirely support the state government and educate every child in the commonwealth.[13]

But the majority report of the Canal Commissioners was succeeded within a few days by a minority report that threw a dash of cold water on all the high hopes. For the third Commissioner, Charles Trcziyulney, took an extremely skeptical view. With remarkable caution for those days, he refused to decide on the most suitable route without a thorough examination of all the proposed routes—a task for which the commissioners had lacked both the time and the experienced engineer. More important, he considered the tunnel proposal entirely impractical: it would be extremely expensive, it would take an enormous amount of time because of the limit on the size of the digging force at any one time, and it required too many locks on each side of the tunnel in too short a space. In his opinion, a canal in the summit area was out of the question: "The whole country . . . is mountainous; mountain rising after mountain in quick succession. . . . Here nature has refused to make her usual kind advances to aid the exertions of man; mountains are thrown together, as if to defy human ingenuity." [14]

The commissioners' reports touched off an intense year-long public discussion. On the one hand, the recommendation of a canal and of a definite route on the basis of very little information aroused the opposition not only of the sections opposed to any canal, but also of the areas on the alternate routes.[15] On the other hand, the Trcziyulney report brought to a head the serious doubts afflicting many of the advocates of a mainline concerning the technical feasibility of a canal over the mountain range. The proposal for a four-and-one-half-mile tunnel was particularly disturbing. Tunnels were relatively new, as long a one as this unheard of. It was natural therefore for many to turn to the railroad as a possible solution. The debate on whether to build a canal at all was from this point on supplemented by a debate within the mainliners' ranks on the relative merits of canals and railroads.

Although there was as yet no railroad in the United States, there was a railroad enthusiast. For a decade and a half John Stevens had been trying to bring the new method before the attention of the public. In 1811 he applied to the New Jersey legislature for a railroad charter; in 1812 he seriously proposed to the New York Board of Commissioners for the Improvement of Inland Navigation that they build a railroad instead of the Erie Canal; and in 1815 he received from New Jersey the first American railroad charter. But he was unable to raise the capital for so daring a venture as a railroad from the Delaware to the Raritan, and his charter lapsed.[16]

In 1820 Stevens turned his restless attention to Pennsylvania. Foreseeing tremendous advantages in a Philadelphia-to-Pittsburgh railroad—he predicted that it would draw the trade of the entire west to Philadelphia—he offered to run a quarter-mile railroad on Philadelphia's Market Street to demonstrate its potentialities.[17] The City Council showed no interest in a Stevens demonstration, but he obtained a somewhat more sympathetic hearing from Philadelphia's businessmen, who were becoming agitated both by Erie Canal construction and by the development of steamboats on the Mississippi River.[18]

In 1822 Stevens applied to the State Legislature for a charter for a Harrisburg-to-Pittsburgh railroad that would connect the Union Canal, then under construction, to the west. But the Legislature, afraid of Baltimore's demonstrated ability to attract Pennsylvania's produce down the Susquehanna, gave him instead, in 1823, a charter for a "Pennsylvania Railroad Company" from Philadelphia to the Susquehanna at Columbia. Stevens managed to attract some influential support for this project, but attracting capital was a different matter. There was general skepticism about a railroad as unprecedentedly long as 73 miles and a disbelief that a locomotive could operate on anything but an absolutely level track. Furthermore, there was serious discussion by this time of the possibility of a canal from Philadelphia to the Ohio.[19] In a long open letter to the public, Stevens attempted to counter the general opinion: a long railroad was in principle no more difficult than a short one, and the proposed line would at no point rise more than two degrees from the horizontal. But he could not raise even the $5,000 necessary to construct a test mile.[20]

During the following two years, however, American attitudes toward the railroad were profoundly influenced by the news of a railroad mania in England. The mania developed as part of a general speculative boom, but it was also based upon the important advances in railroad technology accomplished since the beginning of the century. In rails, the edge-rail had replaced the flat plate, the flange had been transferred from rail to wheel, and malleable iron had begun to replace cast iron. The first locomotive was used in 1804, there was constant improvement in the following two decades, and by the early 1820s Stephenson's locomotives were attracting attention all over England. By 1825, railways connecting all the largest cities were being planned and at least five railway companies, in addition to thirty dock, loan, insurance, and similar types of companies, were being floated. These flotations were mainly speculative—few schemes reached the construction stage—but the building of the Stockton and Darlington Railroad increased the general enthusiasm.

The timing as well as the success of the Stockton and Darlington were extremely important for the development of the Pennsylvania debate. A charter for a horse-drawn railroad carrying only freight was obtained from Parliament in 1821, and George Stephenson was appointed engineer. In 1823, when the roadway was far advanced, the company obtained an amended charter which permitted the use of locomotives and stationary engines for the hauling of both goods and passengers and which gave the right to all to place their wagons on the road in payment of a toll fixed in the Act. The Stockton and Darlington was eminently successful, and an important factor in that success was speed. When it opened, on September 23, 1825, observers were amazed to see a locomotive haul a train weighing at least eighty tons up an incline at ten to fifteen miles per hour. However, the line was not a modern general-purpose railroad in the full sense, for it drew mainly coal, and its passenger traffic developed slowly before 1832.

With the success of the Stockton and Darlington assured, eighteen new railroads were authorized by Parliament in 1826. But three of the most important of these had been projected before the Stockton and Darlington opened. The campaign for the Liverpool and Birmingham began in 1824; there was some agitation for a Birmingham and London Railroad in 1825; and the first prospectus for the Liverpool and Manchester was issued in October, 1824.[21]

All these developments were reflected in the Philadelphia press. The *National Gazette* of January 20, 1825 quoted a petition for a "general iron Rail Way" to the city of London which listed the advantages of railroads and enlarged on "the folly of making canals . . . in the neighborhood of London." The Philadelphia editor was greatly impressed: "The prodigies to be effected with Rail Roads are strikingly exhibited." In another Philadelphia daily, one writer quoted the London *Gentleman's Magazine* on railways and had no doubt of their application to many parts of Pennsylvania, another proposed a railway instead of the Schuylkill-Delaware Canal, and a third described the growing importance of iron railways.[22] In February the State Senate responded to the growth of railroad interest with the appointment of a committee to investigate the possibility of building a railroad from Philadelphia to Pittsburgh.[23]

In the same month, the Pennsylvania Society published a pamphlet on railroads that consisted of a digest of an essay by the well-known British engineer, Robert Stevenson. It went through three editions within six weeks and was reprinted in full in the *United States Gazette* of March 3. Stevenson's explanation of the recent rapid development of railroads in Scotland and Wales must have been of particular interest to his American readers. It was England's great wealth, he explained, that had allowed her to be first in the development of canals and other waterways. Her less wealthy neighbors were now trying to make up for their lack of these improvements by constructing numerous railways, "which are perhaps better adapted than canals to the undulating surface of their respective countries; while they are most economical and more generally applicable to the ordinary purpose of commercial traffic." [24] Thus, everything happening in the British Isles at the time—the technical advances in track and locomotive, the projects for general-purpose railroad lines between the urban centers of England, railroad construction in Scotland and Wales, and above all, the imminent opening of the Stockton and Darlington—spoke in favor of keeping an open mind on the question in Pennsylvania.

All of the railroad interest in the Philadelphia press thus far described was purely informational; it represented no public debate. On February 18, however, the first flash of battle appeared in the *United States Gazette* in the form of an editorial blast against railroad advocacy in Pennsylvania. The editor began with relatively calm references to the lack of railroad data and the need to wait for the report from Europe of the Society's agent, William Strickland—a tranquilizing theme often to be used by the mainline leadership in the succeeding months. But in the meantime, the editor added, it was worth arguing whether the vast advantages of canals were to be entirely given up "because there is a possibility that some more *plausible* undertaking *may* be proposed." The editor was deeply disappointed in the railway advocates: "We had hoped that a majority of our citizens were prepared to make a sacrifice of narrow prejudice and selfish views, upon the altar . . . of public good." The phrase was remarkably bitter; "narrow prejudices and selfish views" was the stock characterization formerly reserved for those who opposed any mainline. But this was only a mild beginning. Before the debate was over, canal advocates were accusing the railroad party of "throwing foreign dust in the public eye." [25]

By March the great debate was in full swing. The railroad party threw down the gauntlet with the publication of an anonymous pamphlet that in its various editions was to become the central point of defense and attack during the subsequent months. This first full statement of the prorailroad position in Pennsylvania judiciously reviewed and answered all the arguments against the railroad and boldly proposed a full railway line from Philadelphia to Pittsburgh. The railroad would utilize the existing bridges over the Susquehanna, employ inclined planes over the Alleghenies, and cost but $3 million. Horses would provide the sole motive power except on the inclined planes, where extra horses, steam engines, or water power might be used. The author confidently heralded the dawn of the railroad age: "May we not confidently anticipate the period when . . . railroads will be known as the only rational means of conveyance?" [26]

The first full-scale review of the route and the railroad questions by a leader of the state's mainline forces appeared in a pseudonymous article by Mathew Carey in the *United States Gazette* of February 16.[27] To Carey, the principal problem posed by these issues was the dissension they aroused in mainliner ranks. "These differences of opinion threaten greatly to retard for the present, or even to defeat, the object we all have in view." And they came at a dangerous time, when New York's enterprise was "gradually undermining the foundations of our prosperity." The mainline's interest demanded a suspension of the argument: "Let us all unite, heart and hand, in the purpose of opening a communication, and let the mode and the route be subjects of future consideration, after proper explorations." More information on railroads was required: "We are much in the dark on this subject." Light would soon be shed by Strickland's reports from Europe. These were expected in July or August, early enough to enable the leaders of the movement to form "sound views" in time.

Having made the plea for a suspension of the debate, Carey went on to give his opinions on the relative advantages of railroads and canals, evidently with the purpose of moderating any dangerously optimistic views of railroads. The argument was couched in judicious terms. Carey saw considerable advantages in railways: the greater speed; the diminished liability to obstruction and delay, especially during the winter; the ease of repair. But there were serious disadvantages: the American frosts, much more intense than those in England; the heavy expense of railroad construction; and, "more than all," the great elevation to be overcome. Carey conceded that the problem of frost might be overcome by sinking the blocks on which the rails rested to a greater depth than was the practice in England and recommended a small-scale experiment to determine the effectiveness of this method. Concerning cost of construction, he quoted an "able and elaborate memorial drawn up at Pittsburgh" which estimated the cost at $8,000 per mile but which did not say whether this was for a single or double set of tracks. Carey conceded that if a complete railway could be built for $8,000 or even $10,000 per mile, "the idea of a canal ought to be abandoned . . . in favor of a grand railway through the whole route." But he was doubtful of the Pittsburgh cost estimate. The Manchester and Liverpool Railroad, he pointed out, was built on relatively level ground, had only a single set of tracks, and had the advantages of English skill and experience and cheap labor. Nevertheless it would cost $54,000 per mile. At that rate, he estimated the cost of a Philadelphia-to-Pittsburgh railroad at $16,200,000—an impossible sum in his opinion. A double railroad would cost fifty percent more. Clearly, Carey was extremely skeptical, but his principal concern was to avoid the damaging effects of the spread of dissension on the issue.

As for the Pennsylvania Society, its publications stressed that it favored no particular route or mode of communication, that its only function was to press for a mainline in general and to find and disseminate facts about internal improvements. There is no reason to doubt that this policy of neutrality reflected the real opinions of the Society's leaders in this period. True, the Society published essays on the advantages of canals and on the necessity and benefits of a water communication between the Susquehanna and the Allegheny.[28] But not only did it publish the highly favorable pamphlet on railroads described above; it also lobbied in the Pennsylvania and New Jersey legislatures during the first half of 1825 on behalf of the railroad ventures of one of its charter members, John Stevens.[29]

The Strickland mission is further evidence of the open-mindedness of the Society's leaders. Their instructions to him regarding railroads stressed the lack of knowledge in the United States upon three subjects: mode of construction, expense, and the maximum angle of ascent on which a railroad would be advantageous. The instructions pointed out that, though railroads were still "subjects of controversy and doubt" in England, Strickland would arrive there at a particularly fortunate time, for "the great communication . . . between Manchester and Liverpool and between Birmingham and Liverpool will have been commenced, or all the principles and plans . . . will have been . . . determined." The Society stressed the need of a prompt report; Strickland was ordered to commence his railroad inquiries before his other assigned subjects and to transmit his information as quickly as possible.[30]

Thus, by March of 1825 the issue between railroad and canal was fully posed but, for the leadership of the mainline movement, not yet joined. For it was understood that action in America could not, as yet, depend on debate in America. "The railroads which are about to be made in England, will furnish the best data upon which to reason and act in this country." [31] Furthermore, the instructions to Strickland indicate that there was general confidence that the data soon to come from England would be sufficient to settle the issue. The principal problem, therefore, was to avoid the development of an all-out battle within the mainline party before the decisive data arrived. Hence the protestations of impartiality and the pleas for a suspension of opinions were oft repeated and more and more urgent. In March the Acting Committee of the Society no longer automatically called for a "canal communication," but for the "opening of a communication by canals (or railways should they appear, on full examination, to be preferable, or by both canals and railways)." And while affirming its impartiality regarding both mode and route, the Committee deprecated "as a most serious calamity" any difference of opinion that might cause dissension among mainliners. The important thing was to concentrate all energies upon obtaining a mainline. "After diligent explorations are made by competent and disinterested individuals, then will be the proper time to determine on the route, and on the comparative merits of canals and railways." [32] Carey's next article even suggested that the bill for the appointment of a new Board of Canal Commissioners should direct the engineers to determine suitable location for railways as well as for canals, because railways "may, on full examination, be found preferable to canals, particularly in mountainous parts of the country." In view of Strickland's mission, Carey was certain that "full and ample information on the subject" would be available in three to five months, "which will be sufficiently early for the purposes of the commissioners." [33]

The leadership's appeal for a few months' postponement of the railroad discussion was successful; the case for delay was strong and Strickland's "full and ample" information was promised very soon. During April, May, and June the railroad as an issue virtually disappeared from the Philadelphia press—only scattered references to English railways remained—and interest again centered upon agitation for a mainline in general. However, Carey's suggestion that the mainline bill should be worded impartially as to mode of transportation was not heeded. The new bill, which became law on April 11, 1825, called for appointment of a board of five canal commissioners, which was to prepare the establishment of a "navigable communication" across the state and to Lake Erie.[34] But the wording of the bill did not matter very much. Action was still in the survey stage, and surveys were not begun until the new board was organized in July.[35]

The April law was a long step forward, but mainliners were still far from their goal of actual appropriations for construction. Further pressure on the Legislature was required. Consequently a committee of twenty-four that had been set up by a Philadelphia mass meeting the previous January to carry on agitation for the mainline issued a call in April for a state-wide internal improvement convention.[36] The convention met at Harrisburg from August 4 to 6 and passed the customary resolution for a "line of communication" by the "best practicable route"—thereby avoiding both the route and the railway issues. It took a strong stand, however, against "diffusive and unconnected application of the public means." [37] The convention then rejected a resolution introduced by Charles J. Ingersoll and William J. Duane of Philadelphia, which recommended a mainline railroad and lateral railroads to the "early and earnest consideration of the constituted authorities of the state." [38]

But the issue of a mainline railroad was far from dead. After three months of relative quiescence, it flared up again in the summer of 1825 and continued unabated for the remainder of the year. The first important stimulus to debate came in July with the publication of *Facts and Arguments in Favour of Adopting Railways in Preference to Canals in the State of Pennsylvania*, a new, much enlarged, and widely circulated edition of the anonymous railroad pamphlet described above.[39] The pamphlet provided a thorough discussion of the technical problems and advantages of the railroad, together with some acute comments on the future of the railroad for the world, as well as its importance for Pennsylvania. According to Carey, it was "circulated through the state with great zeal and industry." [40] Judging from the number of times it was referred to during the debate, it became the basic reference work for both the pro- and anti-railway forces.

The author of *Facts and Arguments* pointed out that there was every reason to expect a continuation of the rapid technical development of the railway: "Various improvements are daily discovered in the construction and management of railways, and the vehicles used on them. . . . Additional improvements may be confidently anticipated." Prevention of the oxidation of iron, lighter and stronger cars, a gas vacuum engine to supersede the steam engine, smaller axles, and other developments were mentioned as possible or probable. In view of these potentialities, "it does not require the voice of prophecy to predict that . . . the New York canal will be superseded by a railway." But before that event it would become evident that Pennsylvania's delay in effecting a communication between the Ohio and the Delaware, so often regretted by the "enlightened citizens of Philadelphia," was really a blessing in disguise. It had prevented a loss of capital on an obsolete canal and thus made possible a railroad that would "monopolize the commerce of the Western country." For the route from New York to Pittsburgh via the Erie Canal, Lake Erie, and the Allegheny River required four transshipments and was frozen five months of the year, while the railway route from Philadelphia would require no transshipment, would operate all year, and would be half as long. Furthermore, while the interregional railroad defeated the "foreign" enemy, the local railway would solve the vexing sectional problem. Railroads could be built so cheaply compared to canals that all the routes demanded by all the sections could be built for an amount not exceeding the cost of a canal mainline.[41]

The pamphlet had a remarkable though temporary effect upon the policy of the *United States Gazette*. The paper had opposed railroad advocacy in Pennsylvania and, two weeks before publication of *Facts and Arguments,* had vigorously defended from its detractors the February majority report of the Canal Commissioners.[42] But a reading of the pamphlet led the editor to advocate a reexamination of the problem: "Whatever may have been our prejudices in favor of Canals, as being the most splendid, and withall the most fashionable mode of improvement," it was necessary now to lay aside such "prepossession." Canals, he pointed out, had been favored because of their success in New York, while railroads had not yet been tried in the United States. England, however, was beginning to erect railroads side by side with the most profitable canals, a proof of the opinion of the well-informed there. Furthermore, the pamphlet had refuted, to the editor's satisfaction, the widespread belief that the mountains presented an insuperable obstacle to the railroad. The editor was aware that "practice is to be regarded as preferable to theory." But practical demonstrations of the railroad were imminent; delay was therefore advisable. The Legislature should resolve immediately in favor of a mainline, but the mode "either by Rail Road or Canal, or perhaps by both" should be decided in the light of further information. The editorial ended with a reference to a crucial source of that information: "It is understood that [Strickland] . . . will report favorably to Rail Roads."[43] The next day the paper attacked the antimainline editor of the *Miner's Journal* for attaching an "undue importance" to canals: "He appears not to have possessed himself of the most recent reports of gentlemen of undoubted science, upon the value of railroads."

This growing prorailway sentiment was given a further great impetus when the crucial section of Strickland's first report on railroads was published in the Philadelphia press on August 12.[44] This was the information so eagerly awaited by the mainline leadership—the information that would settle the issue between railroad and canal. It was the expectation of this report that had justified the mainline leadership's successful attempt to postpone the divisive railroad debate. In view of its central importance, it is worth quoting in full:

As to the relative advantages of railways and canals in mountainous or level countries, there appears to be but one opinion among the ablest Engineers in England; both modes of transportation have been practically tested, and although much wealth and commercial greatness have been produced by numerous canals, still railroads offer greater facilities for the conveyance of goods, with more *safety, speed,* and *economy.*

It is a matter of little importance whether the surface of the country where they are introduced be level or mountainous; these objects are still attained without difficulty; if *level,* horses or locomotive engines are to be used with great advantage; if *mountainous,* a stationary steam-engine or brake is to be applied on the summit; while the ascent is to be overcome at once, by means of an inclined plane. It would not perhaps be proper for me to eulogize this system of internal improvements in a report which is solely intended to convey facts; but I feel it a duty which I owe to my own judgment, as well as to that of the liberal Society whom I have the honor to serve, to state distinctly my full conviction of the *utility* and decided *superiority* of *Railways* over other modes as a means of conveyance; and one which ought to command the serious attention and adoption by the people of Pennsylvania.

The publication of this report marked the end of whatever remained of the railroad truce. Instead, the Philadelphia press developed an enthusiastic interest in railroads, an interest that was fed by the news of the railroad mania in England. The following items, appearing in Philadelphia newspapers during the fall and winter of 1825, will illustrate the trend. In September, plans were reported for the exhibition of a 1,000-yard railway and a locomotive on Capitol Hill in Harrisburg for the benefit of Pennsylvania's legislators.[45] In October a public meeting in the Borough of Columbia called for the immediate construction of a Philadelphia to Columbia railroad.[46] In November came the impressive news that the continent was falling into step with the British Isles. Strickland reported plans for an 80-mile railroad between Hanover and Hamburg, while another source reported the building of a 280-mile iron railroad in Poland between Kalish and Brezec.[47] Furthermore, the locomotive was proving itself. From London came a detailed account of the powers and advantages of the kind of locomotive used so successfully on the Stockton and Darlington. The locomotive traveled "with ease and safety with a weight of 90 tons in its train, at the rate of eight miles an hour" and thereby proved the great advantages of the railroad over the canal.[48]

Thus the leaders of the mainline movement were confronted with a rapidly developing railroad enthusiasm. According to its own later account, the response of the Acting Committee of the Society to Strickland's first report was also favorable. The committee praised the comprehensiveness of the report, for it provided every bit of information required to introduce railroads into the United States and "to establish their competency for the purpose to which they are applicable . . . " Strickland also procured for the Society a working model of a locomotive engine. "A machine so valuable, of such astonishing competency," the committee believed, "ought to be more generally known in our country." [49]

Nevertheless, the committee was disturbed by the rapid growth of railroad interest and decided that it must quickly obtain from Strickland a more detailed, and possibly a more qualified, opinion on the merits of railroads. Its letter to him of September 19 anxiously referred to the excitement in the state on the railroad issue and enlarged on its possible effects: "It is feared by many, that . . . it will divide the friends of the cause of improvement, and thus postpone, if not prevent the commencement of the great work." [50] The committee therefore demanded certainty; Strickland was warned that unless he furnished "facts and arguments of an entirely conclusive nature" in favor of railways, Pennsylvanians "will pause," for they did not have his opportunities of testing the railroad's advantages. A series of questions followed: Did Strickland contemplate using railroads for both goods and passengers, and, if so, was not a double railroad needed for each purpose? Did he contemplate the use of locomotives on railroads in Pennsylvania? Had he considered the difficulties of using steam engines on inclines, and in the midst of mountains far from the skilled engineers needed to repair them? Would not a railroad cost at least $20,000 per mile if iron cost $35 per ton? Would not the demand for iron for a railroad from Philadelphia to Pittsburgh increase the price? Could present establishments in the United States provide sufficient iron within a reasonable time? The questions indicate serious doubt but also serious consideration of the railroad:

"These propositions are not suggested as insurmountable difficulties, but as specimens of the matters which will be required of you when you return." Indeed, the committee ended with a reassertion of its impartiality: "It is not for the Pennsylvania Society to adopt a preference for any particular plan of improvements, or to discourage investigation. Its object is the improvement of the state by the best plan."

In his answering letter from Liverpool, dated October 20, Strickland politely rejected the committee's implication that his opinions were contributing to a dangerous railroad-canal division. As to the specific questions, he explained that two pairs of tracks—which, he emphasized, had to be edge rails—would be sufficient. But, he added, they must be constructed for use by locomotives, for only with these did the railroad have a decisive superiority over the canal. Strickland was convinced of the "practical efficiency" of the locomotive on moderate inclines as well as on levels. For steeper grades he advocated the current English practice of inclined planes and stationary engines. Locomotives, he thought, would be much more efficient than were horses on canals; and stationary engines and inclined planes were much cheaper to construct than were canal locks, while their maintenance and repair were not more difficult. He admitted, however, that railroad construction would force up iron prices and that the inability of the state's industry to supply sufficient iron within a reasonable time would be a great disadvantage.

All these specific arguments were important; but Strickland's concluding remarks took up a more basic source of the committee's anxieties. The railway in England, he conceded, had been satisfactorily proven on a relatively small scale, and for limited purposes. True, it had succeeded on "varied surfaces of country," and Englishmen "of intelligence and capital" were confident of its general extension. But he acknowledged that his "honest opinion derived from a personal investigation of *facts* [which] . . . may be presumed to have been exhibited, perhaps upon too small a scale in England, to admit of an unequivocal recommendation to your society, or the people of Pennsylvania."

A week after the Acting Committee of the Society had addressed its anxious letter to Strickland, its chairman, Mathew Carey, launched an all-out campaign against a railroad mainline. His action must have been a complete surprise to many mainliners. As chairman of the Acting Committee, Carey had emphasized its impartiality.[51] As an individual writing under his well-known pseudonyms of "Fulton" and "Hamilton," he had professed neutrality while counseling caution until the facts were unearthed by the Strickland mission. So uncommitted did he seem at the time in both his personal and his official capacities that, despite a skepticism toward the railroad that is evident in his writings from the beginning, the *United States Gazette* could mistakenly assume that the Society had published the best-known document of the prorailroad party, *Facts and Arguments;* an antirailway debater could complain that the pamphlet was "decidedly patronized by the society"; and another writer could assume that Carey had written the pamphlet.[52] Another factor likely to give the Society a prorailway tinge was Carey's early recognition of the great technical difficulties of building a canal in the mountain area. Consequently he hinted at the possibility of a railroad segment to connect the canals on each side of the mountains. Conservative opponents of the railroad—those who opposed all railways as so much nonsense—would hardly distinguish between segments and long-range lines and might well have considered both the fact-gathering activities of the Society and the suggestions of a railroad segment as cut from the same visionary cloth.

But Carey now took a definite position. In a series of articles written between September, 1825, and February, 1826, he launched an all-out campaign against railroad advocacy precisely at the time when railroad interest in Philadelphia was reaching fever height.[53] The importance of Carey's actions can hardly be overestimated. He possessed by far the best-known name and the most skilled pen in Pennsylvania and he had sufficient means to devote all his time to public activities.[54] In addition, as founder and vice-president of the Society and as chairman of the Acting Committee that performed the Society's day-to-day work, he was in the best of positions to influence events. His influence was such that a prorailroad opponent felt compelled to warn that, when such a man entered the lists, there was danger of bowing "to the blandishments of reputation more than to the strength of argument." [55]

Carey was not alone in his attack. His pamphlets were accompanied by a stream of bitterly antirailroad literature from various pens, provoking from the railroaders replies equally bitter. This remarkable public discussion explored every technical and economic aspect of the choice before Pennsylvania. For half a year, a flood of resolutions, pamphlets, editorials, and letters to the newspapers tested the public's patience for extended technical analysis, for widely variant statistics, and for repeated rebuttal and counterrebuttal.

The railroad advocates vigorously defended the practicability and advantages of the railroad. They pointed out that as a result of its independence of a water supply a railroad could be constructed anywhere; its capacity could be expanded at will, simply by adding tracks; shorter, more direct lines could be followed; goods would be less subject to damage; year-round operation would be feasible. Then there was the ability to increase speed indefinitely at very low extra cost—in their opinion the decisive advantage over the canal. This required that the locomotive replace the horse, but they could point to the general acceptance of the locomotive in England and to Strickland's report that its practicability had been fully proved. Nevertheless, the author of the principal prorailroad work, *Facts and Arguments,* suggested that the choice between horse and locomotive be postponed until more information from England was available.

The locomotive of those days, however, posed one serious problem: it could negotiate only moderate inclines. This limitation had hindered the development of the general-purpose railroad in England for a decade or two. On the typical mine railway, freight moved down the incline only; the cars were empty on the way up. If a railroad for general two-way traffic used locomotives, it was limited to relatively level areas. But Strickland reported that a solution had been found: a series of road segments were made level or limited to moderate inclines for as long as possible by means of tunneling, cutting, and winding around hills. These could use locomotives. The segments were connected by inclined planes each fitted with stationary engines at the top to draw up and let down the cars.

These were the advantages of the railroad. Although some of them were minimized by the canal party, they could not be entirely gainsaid. However, there was one serious disadvantage of the railroad method. The use of tracks imposed a rigidity that did not exist on canals and roads; it was impossible for trains traveling at different speeds to pass one another. Since it was taken for granted that most shippers would own and control the cars that carried their goods, chaotic conditions could easily develop. Two or more sets of tracks in each direction would be required, the canal party asserted, and this would vastly increase the expense of a railroad. However, William Strickland had reported to the Pennsylvania Society that in England the problem was solved by means of periodic sidings or, in the case of double-tracked railroads, by switches.

A final issue that took up a good deal of the time of the disputants was the cost of a Pennsylvania railroad. On this point everyone was in the dark, but no one would admit it. British railroad costs were quoted, rebutted, and quoted again, but even if the combatants had agreed as to the British statistics, they could have debated endlessly both on their applicability to the very different cost conditions in the United States and on the possibilities of adapting modes of construction to American conditions. The author of *Facts and Arguments* estimated the cost of a double-tracked horse railway between Philadelphia and Pittsburgh at $3 million; a canal, he thought, would cost at least $7 million, and probably double that amount in view of the fact that the Erie, with so much less lockage and no tunnel, cost $7 million plus interest. Carey believed that the cost of railroad and canal would be about equal.

None of these issues seem crucial; certainly, there was no decisive technical objection to the railroad that would explain Carey's sudden attack upon it. Indeed, in a limited area the canal advocates were actually forced to accept the railroad despite their criticisms. During the debate it gradually became clear to the principal mainline leaders that a full canal was out of the question, not only because of the expense and time required by a tunnel under the summit of the Alleghenies, but also because the water supply would be insufficient even at the tunnel level. Consequently, during the later stages of the debate they stopped referring to an all-water route; while continuing to oppose construction of a full railroad, they were prepared to accept a railroad portage over the mountains.

Acceptance of the portage railroad directly contradicted a number of the technical arguments that canallers advanced against the full railroad. A full railroad would need several sets of tracks, they thought, but the portage railroad would need but two. The severe American frosts would break up the tracks of railroads in general, but not of a portage railroad in a high mountain area. That locomotives could not be used on steep inclines was considered a serious objection to the railroad, but the portage railroad could use to good effect the system used in England for entire railroad lines and recommended by Strickland, that is, locomotives on relatively level areas and stationary engines on inclined planes. It seems evident that these technical issues did not determine the choice between railroad and canal.[56]

Another extremely interesting aspect of the debate was the clarity of the railroad advocates' vision of the future of the railroad. They perceived that the railroad was the historic solution to the long-range transportation problem of continental areas. This inspiring insight was probably the railroad party's most effective counter to the wave of canal enthusiasm generated by the Erie Canal. For they knew, as we know now, that the potentialities of the railroad were such as to put even the Erie Canal into the shade.

It was the speed factor especially that fired the imaginations of the railroad enthusiasts. For they had come to believe that there was "scarcely any limit" to the velocity; that it could reach such heights that air resistance, ignored up to that time, "would become the principal retardative force." The social consequences of this new mode of transportation were pointed out in a perceptive British article reprinted in Philadelphia. The outlying towns of an extensive empire, the author predicted, would be transformed into "so many suburbs of the metropolis." With commodities, inventions, opinions, and feelings circulating with enormous rapidity, with "the personal intercourse of man" vastly increased, he even anticipated the unification of Europe as the result of the building of a railroad system. As for America, "the 150 million who will inhabit North America next century will be more completely *one people* than the inhabitants of France or Britain at this day."[57]

These would be the long-run, universal effects of the railroad. Concerning its implications for the more immediate and mundane problems of Pennsylvania, the railroad advocates were equally accurate. Firstly, they realized that the railroad would have a decisive effect upon the rivalry with New York City: "Those towns on the seaboard of the United States that cannot have an extensive intercourse with the interior by Canals, will be compelled to adopt Rail Roads."[58] Secondly, they forecast the consequences for Pennsylvania if a canal mainline were built. Sometime in the near future, a railroad advocate predicted, "public" general purpose railways would be proved by experience to be as superior to canals as were "private" specialized mine railways. At that time, "some of our enterprising capitalists" would find it profitable to build a full railroad from Philadelphia to Pittsburgh in competition with the canal mainline. Such a railroad, he asserted, would destroy the value of the canal. It would carry freight more cheaply, with greater speed and safety, by a shorter route, and throughout the year. "Should the advocates of canals gain a victory," he warned, ". . . *it is one they will never boast of*. An error committed at present will have the most serious consequences on the future destinies of this state."[59]

In December of 1825 the scene of battle shifted to the legislative halls. Governor John A. Shulze's annual message to the Legislature reflected the widespread doubts of a canal policy.

Describing the great natural obstructions to mainline construction—obstructions that did not exist in New York—he emphasized "the advantages to be gained by waiting for full and perfect information before any irretrievable steps should be taken." [60]

But those "irretrievable steps" were recommended to the Legislature the following February in the report of the new Board of Canal Commissioners (appointed the previous April). The Board rejected two of the previous Board's recommendations: it proposed a shorter route and suggested a railroad portage instead of a tunnel. But the work could be started without settling these issues immediately; the Board urged the legislature to proceed with those canal sections that were not in the mountains and that were common to all the routes under consideration. These consisted of a canal along the Susquehanna between Middletown and the mouth of the Juniata and another along the Allegheny between Pittsburgh and Kiskiminetas Creek. The Union Canal, which was scheduled to begin operation in the spring of 1827, would temporarily connect Philadelphia to the mainline. It was assumed that a larger, more direct canal would be built in that area sometime in the future.[61]

The Legislature agreed with the Canal Commissioners; a bill for the construction of fifty-four miles of canal at both extremities of the mainline route passed by a vote of 19 to 14 in the Senate, 65 to 23 in the House. It was signed by the Governor on February 25. The route issue and the portage problem were left open. To the criticism that it was premature to begin construction of segments before the entire project had been decided upon, the House Committee on Inland Navigation and Internal Improvement replied that the segments would be valuable as independent canals. Their utility would outweigh their cost, the Committee asserted, "even if no extension of the line of canal should ever take place." [62] The fact that the construction of canal segments ruled out the full railroad was not mentioned. Construction was begun on July 4, 1826. Governor Shulze's annual message in the following December reviewed the various improvements in progress without referring to the recommendation for postponement made in his message of the preceding year.⁶³ The railway-canal debate was over.⁶⁴

Why was the railroad rejected? Certainly not for lack of access to the available information. The railroad party accurately foresaw and clearly presented the advantages and potentialities of the railroad, and the canal party raised no definitive technical objection to the method. Indeed, the leaders of the improvement movement turned against the railroad soon after their agent in Europe had reported to them that it was entirely practicable and far superior to the canal. A good part of the explanation for this behavior lies in the Society's correspondence with Strickland. The improvement leaders had fully expected that Strickland's report would settle the railroad issue one way or the other. But when his first report supported the railroad with what they regarded as insufficient proof and they wrote to him demanding "facts and arguments of an entirely conclusive nature," all he could answer was that he could provide no certainty because no long-range general-purpose railroad had as yet been built.⁶⁵ A delay, therefore, was unavoidable. A few years would be needed for more adequate experience with the general-purpose railroad in England and for some experimentation in the different American conditions. Since the Pennsylvanians thought that the situation was too urgent for any delay, they chose the canal method. The cause, then, was the intensity of their reaction to the success of the Erie Canal. There was what may justifiably be termed a panic reaction. "The delay of a year may prove fatal," wrote the Pittsburgh *Gazette*; ⁶⁶ "We now contend for existence, or all that makes existence desirable," warned the *United States Gazette* in regard to Pennsylvania's delays.⁶⁷ This nervous attitude was displayed even toward the Strickland mission to Europe. The Society expected his report within a few months; Carey, in March, expected "full and ample information" in three to five months.⁶⁸ But even this short delay provoked a sharp editorial protest from the *United States Gazette:* "We fear that the love of procrastination, that 'sin which most easily besets us,' will paralyze . . . improvements." ⁶⁹

The changing attitudes of the *United States Gazette* toward the railroad illustrates the importance of this fear of delay: As noted above, this leading Philadelphia daily had at first reacted with hostility to the railroad idea, but the publication of *Facts and Arguments* during the summer made a great impression on the editor. Subsequently the paper urged serious consideration of a full railroad line.[70] By October, however, the paper was stressing, not the relative advantages of railroads and canals, but the terrible dangers of delaying the mainline: "It is necessary to our prosperity, nay, almost to our existence, and every day that the work is delayed, is strengthening and multiplying the advantages that our gigantic neighbor has acquired over us." [71] Consequently the paper was entirely in favor of the mainline canal construction bill of February, 1826. The bill was passed in a great hurry—before adequate surveys could determine the route and before the method of crossing the mountains was decided—and the *Gazette* did admit that, in passing the bill, the House of Representatives might "appear to some rather hasty." But haste was an absolute necessity: "By further delay, the public *excitability* would have been wasted, and so powerful an opposition raised to the measure, that its final adoption would have been doubtful." [72]

Carey's remarks on the fate of the Liverpool and Manchester Railroad bill in the committee of the House of Commons provide the clearest illustration of this attitude. The bill failed by a majority of one, and Carey utilized this fact in arguing that English opinion was not unanimously on the side of railroads.[73] He added, however, that he had earnestly hoped that the bill would pass and "thus, a fair experiment been early made with foreign capital, so as to enable the state of Pennsylvania to profit by that experiment, before she decided the important question between canals and railways. The failure of the bill, therefore, is sincerely to be regretted." [74] But the bill was to be reintroduced at the next session of Parliament. Even a short delay was evidently out of the question.

The symptoms of panic were particularly pronounced whenever the Erie Canal was mentioned. "Whenever we hear of the

New York canal a kind of tremor seizes our frames. Whenever we hear Mr. Clinton's speeches or reports, we can hear old Cato's *'delenda est Carthage'* thundering in our ears." [75]

It is a striking fact that the Erie Canal was not considered as simply an aid to a competitor's trade; the fear was not that New York, already larger than Philadelphia, would be still larger in the future. Such a prospect would have been considered regrettable but hardly catastrophic—Philadelphia would also have grown, if not as rapidly. Instead, the improvement advocates wrote as if the Erie Canal confronted Philadelphia with utter stagnation. To the Committee of the Harrisburg Convention, the issue was, "whether we shall allow our state to be drained of its wealth, to increase that of a neighbor already advanced to a towering height." [76] As a bitterly antirailroad mainliner put it:

We observe our population daily diverging from our state. . . . We have beheld our markets neglected . . . we observe our stage coaches pass empty; and our roads with hardly a solitary traveller; while we hear that 40 stages leave Albany daily, with 700 passengers. We have found that while we slept our *locks* have been shorn.[77]

A part of the explanation for this remarkably intense reaction is that the Erie Canal was considered the symbol of all the projects that were soon to surround Philadelphia. As one Philadelphian put it, "A most important reason for acting definitively at this session is that Ohio will certainly commence her Canal next summer." If it were finished before the Pennsylvania canal, New York would have an "irretrievable advantage" in the Lakes trade. On the other side, he warned, Maryland and Virginia were about to commence a Potomac canal to the west, for the United States Engineers had reported in favor of it. If these three canals—the Erie, the Potomac, and the Ohio—were all completed before Pennsylvania's, they would draw the commerce, the wealth, and the population of Pennsylvania, "and leave us to weep over our inertion [sic] and infatuation." [78]

Another element that helps to explain the extreme sense of urgency in Pennsylvania was the belief prevailing among the leaders of the improvement movement that if they delayed they would lose an opportunity that might never recur. They believed that it was the immediate drama, the spectacular success, of the Erie Canal that had provided them with the opportunity to overcome the great power of antimainline sentiment in the state. "The people are ripe for it," wrote a Philadelphian who opposed railways, "there is a general excitement; and, if this be suffered to expire, no one can tell when so favorable an opportunity will again present itself." [79] Pennsylvania's long record of postponement not only of any serious consideration of a line to the west but of other canal projects as well seemed to bear out this view. Carey later pointed out that even the Chesapeake and Delaware Canal project—much less costly and risky than the mainline— had long been dormant. "At length the noble example set by the Erie and Hudson Canal, aroused the energies of our citizens, which had been so long, as regarded canals, entirely torpid." [80]

Consequently the image of Pennsylvania as a sleeping Samson—"while we slept our locks have been shorn"—had some justification. Another canaller explained railroad agitation in Pennsylvania in this fashion: Assume, he suggested, that a man who has slept long and deeply is suddenly awakened to a situation of great danger. Such a man, in desperation, would be liable to great error in his choice of means. This was Pennsylvania's case, he believed; this explained advocacy of so strange a thing as a railroad.[81] The image was apropos; to the present reader it could better explain advocacy of so strange a thing as a canal over mountains.

Thus, it was the strength of the opposition to the mainline—a strength that in the mainliners' view had crippled the state improvement movement for years and could easily do so again—that evidently explains the now-or-never attitude of the mainline leaders. It was this attitude, together with the great fear of the projects of the competing states, that accounts for the fact that the railroad was considered so dangerous a threat to mainline aims. For the mainline was considered to hang by a thread, but the railroad issue was divisive; and delay meant the end of any mainline project, but serious consideration of the railroad inevitably involved delay. Carey remarked that there was "so much opposition from so many quarters, and on such a variety of grounds" against a mainline that this additional issue "will afford incalculable advantages to those who oppose this grand measure in every shape and form." [82] Other improvement advocates agreed. "If there is as much difference of opinion, and as strongly contested, when the subject is brought before the Legislature," warned "Amicus," ". . . I fear we will be disappointed in seeing either a railroad or a canal for a long time. . . . Amidst all the obstacles and difficulties it will have to encounter in the Legislature, this will most probably be the worst." The great question was *"whether a grand plan of internal improvements shall not be commenced forthwith."* [83]

This attitude had little to do with the much-debated technical characteristics of the railroad as compared to the canal. "I am not by any means an opponent of railroads or of any other improvement that the progressive lights of the age may suggest," wrote a Philadelphian. But, "if at this moment we permit our attention, or the public opinion, to be distracted to railroads, or any other supposed improvements, we shall fail in both plans." The important thing was speed, "before the commerce has acquired the correspondences, and habitudes that are so difficult to break." [84]

This belief in the strength of the opposition to the mainline had another important consequence, one that has dramatically distinguished Pennsylvania's improvement policies from those of Massachusetts and Maryland. It forced the leaders of the improvement movement to drop their announced policy of concentrating all energies on the long-range east-west line. They promised branch lines in return for mainline votes and in building those branches considerably delayed the completion of the mainline. In the east, canals totaling 273.5 miles were built north from the mainline along the Susquehanna and its north and west branches, and a 60-mile Delaware Division connected Bristol (north of Philadelphia) to the coal trade of the Lehigh valley. In the west, 112 miles of canal, aided by a 50-mile canal feeder to provide more water, connected Beaver on the Ohio above Pittsburgh to Lake Erie.[85] The leaders of the improvement movement regretted the effects of this policy but, as with the refusal to postpone the

1826 decision, believed that any other policy would end all hope of a mainline for Pennsylvania. Carey, for example, stated in 1827 that the policy was "the only one that afforded a reasonable prospect of success—that is, consulting the interests of the different great sections of the state, and letting them advance *pari passu*. . . . Any other system would be as impolitic as unjust."[86] Two and one-half years later Carey referred to the simultaneous construction of branch lines and mainline as one of the principal causes of Pennsylvania's financial difficulties.[87] But while regretting the results of the policy, he continued to justify its necessity: the minority opposed to the mainline was so large that alienation of only a few votes would have wrecked the project.[88]

In the mainline leaders' view, therefore, they had little, if any, room for maneuver. The strength of the opposition made it impossible to postpone construction and therefore ruled out the full railroad; and the strength of the opposition necessitated the branch-line deals. This view assumed that the opposition would not be influenced by the technical merits and demerits of the mainline proposal because it was motivated solely by sectional interest and by prejudice against all improvements. Historians have tended to accept this view; they have implicitly assumed that there was no question of poor leadership involved because the leadership had no choice. Both Avard L. Bishop in his "State Works of Pennsylvania" and Louis Hartz in his *Economic Policy and Democratic Thought: Pennsylvania, 1776–1860* completely ignored the railroad-canal debate of 1825. This eliminated one important choice. And Hartz, in stressing the contradictions inherent in the theory of regional self-interest, has given the impression that, aside from prejudice against improvements in general, it was sectional and vested interests alone that determined the strength of the opposition to the mainline proposal.[89]

There is evidence, however, for a different view. If the strength of antimainline sentiment was determined not only by prejudice and sectional interests but also by the real disadvantages of the proposed canal over the mountains, then postponement and a subsequent proposal of a full railroad might well have mitigated the opposition to the mainline. A stronger mainline movement might then have been able to avoid the branch-line policy. The question is worth investigating; certainly the opinions of the mainline leaders as to the sources of antimainline sentiment cannot be taken for granted as accurate. The fact that the failure of the mainline was a result of its natural disadvantages—a fact that Hartz has noted—should in itself cast doubt on the contention that only prejudice and sectional interest caused the opposition to the mainline. For these disadvantages were clearly visible in 1825, and it is difficult to believe that opinion in Pennsylvania was not strongly influenced by these overwhelming geographic factors.

Most of the evidence for this influence has to be extricated from antimainline pronouncements that indicate a complex mixture of motives. Two long antimainline speeches at the Harrisburg Convention raised the following arguments against proceeding with any mainline: (1) there was insufficient evidence of the practicability of the project—four routes had been proposed for the canals, but not one had as yet been sufficiently explored; (2) the estimated cost of tunneling through the summit was ludicrously small; (3) the expense of the canal was uncertain, but it would certainly not be less than $10 million; (4) the finances of the state would not allow so expensive a project—they were against borrowing money for risky experiments; (5) the majority of the population either opposed the project or were lukewarm toward it.

Despite all these objections, the speakers urged, not the scrapping, but the postponement of the mainline. Postponement, they claimed, would provide time to gather sufficient information on the routes, thereby reassuring the population. Furthermore, the people would be influenced in favor of a mainline if the Union Canal, scheduled to open in the summer of 1827, should prove a success. Finally, postponement could do no harm because surveys, estimates, and plans would require eighteen months to two years in any case. They favored continuation of the current Board of Canal Commissioners for these purposes.[90]

Clearly the motives were mixed. The conservative financial

argument against any large project was mixed with specific arguments applicable only to the particular mainline proposal. In the midst of citing all the above arguments, one of the speakers, James M. Porter, interjected that once all the routes were thoroughly examined, "he would go heart and hand with the gentlemen who advocated a grand canal." The other speaker, N. H. Loring, added to the above arguments a panegyric of internal improvements:

Blind must he be indeed . . . who does not perceive the immense wealth . . . which a successful prosecution of the canal policy will give to this commonwealth. . . . On this subject we should not solely calculate the profit and loss. . . . Canals and roads . . . will be the arteries of our political existence.

Thus, not all of the opponents of the mainline were the blind reactionaries portrayed by their more literate and influential Philadelphia opponents; nor were they all motivated by local interest. On the contrary, the real problems involved in a mainline canal, and the haste and lack of clarification of technical issues, made many conservatives wary of a venture which in other respects may have appealed to them. The realism of the antimainliners is sometimes striking; no more accurate contemporary analysis of the real factors involved in Pennsylvania's attempt to compete with the Erie Canal was produced than the following antimainline pronouncement:

The advocates of a grand canal in this state have, in taking the New York canal as the basis of their calculations, entirely overlooked its peculiar advantages. The Clinton canal . . . traverses a country so level that the amount of its lockage does not much exceed the height of Lake Erie above tide water—passes at right angles to the course of numerous rivers that flow from the south, is consequently easily and abundantly supplied with water—possesses along its whole extent a fine wheat country—terminates in Lake Erie, and thus connects an immense inland navigation with the ocean at the city of New York, the commercial depot of America. A canal through Pennsylvania would have nothing in common with this, excepting its termination in Lake Erie. How far it might compete with others for the trade of the lake, may in some measure be estimated by the fact that before it could advance fifteen miles from the lake, it would require a lockage equal to the whole of that of the New York canal.

Mainline propaganda attributed such views to narrow prejudice and selfish interest. But so long as it was the canal and not the railroad that was under discussion, the author's concluding statement was unquestionably correct:

> Our state possesses many advantages—let us improve them. We will certainly fail to compete with the State of New York for the trade of the West. Nature has given her advantages in such a competition which we cannot overcome.[91]

The Pennsylvania decision, then, stands in striking contrast with its New York counterpart. The detailed surveys, careful cost estimates, and full knowledge of the techniques to be used in each area that preceded the Erie Canal decision minimized opposition; in the end, even Martin Van Buren and a good section of the state Democratic party were won over. The Pennsylvania decision, on the other hand, was taken on the basis of sketchy surveys, wildly optimistic cost estimates, and confusion as to much of the route to be taken and as to the method of crossing the mountains. As noted above, the bill for construction of the mainline which passed in February, 1826, authorized construction only of those canal sections that were not in the mountains and that were common to all the routes under consideration. The route issue, the portage problem, and the need to replace the Union Canal between Philadelphia and the Susquehanna were all left for future decision. This extreme haste not only increased the opposition to the project; it also created problems that were to plague the state throughout the period of construction.

The replacement of the Union Canal was considered imperative; mainliners were convinced that it would be a serious bottleneck when the western produce on the mainline together with the Susquehanna trade all demanded passage to Philadelphia— or went downstream to Baltimore. Since the practicability of a larger canal in the area was questionable, the Act of April 9, 1827, provided for a survey and cost estimate for a railroad "with locomotives or stationary engines" from Philadelphia to the eastern end of the mainline canal.[92] The Canal Commissioners reported in the following December that Major John

Wilson of the U.S. Topographical Engineers had advised them that the area between Philadelphia and the Susquehanna was much more appropriate for a railroad than for a canal. Consequently, in March, 1828, the Legislature made the first appropriation for the Philadelphia-Columbia Railroad. It opened in March, 1834, with one track; a second track was completed in October of that year.

The portage problem proved to be far more troublesome. Surveys during 1826 further convinced the Canal Commissioners that a portage was necessary, but, in addition to disagreement as to its form—road, rail, or a combination of the two—there was great public pressure for a continuous waterway. Consequently, the act of April, 1827, required the Canal Commissioners to investigate still again the practicability of a full canal. Meanwhile canal construction continued. The same act provided for the remaining canal sections of the mainline and made the first appropriations for the many branch lines whose construction was to delay the mainline's completion.

But the portage issue remained unsettled. In their December, 1827, report the Canal Commissioners had used the strongest possible terms. They felt "compelled to say, in the most explicit manner" that a continuous canal was "utterly impracticable." But the Legislature would not act. In December, 1828, a new chief engineer, Moncure Robinson, was appointed and instructed to investigate the possibility of building both a railway and a road portage over the mountains. For the Canal Commissioners felt, even at this late date, that a railroad could not handle passenger traffic. Robinson reported in November, 1829, in favor of a railroad portage using inclined planes, but he complicated things still further by reintroducing the tunnel proposal—this time a railroad tunnel, thus eliminating the problem of a water supply.

By this time the situation had become desperate. Construction had been in progress for over three years, survey after survey had been made, and there was still no agreement on the method of crossing the mountains. All the other parts of the mainline were nearing completion but could be used only for local traffic.

There was widespread criticism of the Canal Commissioners. As a result, in March, 1830, the Legislature provided for still another survey of the mountain area!

The engineers reported once again in the fall of 1830. They strongly recommended a railway portage but ruled against a railroad tunnel. On March 21, 1831, five years after work on the mainline was first authorized, the Legislature authorized the building of a railroad portage. Construction began in 1831 and was completed with a single track in March, 1834. During 1835 the second track was completed, and the first locomotive was put in use. The entire cost of the road and all fixed equipment was $1,860,753.

The portage railroad made quite an impression on contemporaries; it was often referred to as one of the wonders of the modern world. It was certainly a pioneering engineering job. Extending 36.7 miles, from Hollidaysburg in the east to Johnstown in the west, it contained ten straight inclined planes, five on each side of the summit, with angles of inclination ranging from 4°9′ to 5°51′. At the head of each plane were two stationary engines of 30 to 35 horsepower each. The road's summit was 1,398.7 feet above the basin at Hollidaysburg, 1,171.6 feet above the basin at Johnstown.

With completion of the portage and the Philadelphia-Columbia Railroad in 1834, the form of the mainline was complete. It was 395 miles long, reached a summit of 2,322 feet above the sea, and cost $12,106,788. In contrast, the Erie Canal was 363 miles long, its summit was 650 feet above sea level, and it cost $7,143,790. There were 174 locks on the canals of the mainline for a total lockage of 1,141.5 feet, in addition to a rise and fall on the Allegheny Portage Railroad of 2,570 feet, making a total rise and fall between Pittsburgh and the Susquehanna of 3,711.5 feet. On the Erie there were 84 locks for a lockage of 689 feet. The prism of the canals on the mainline was the same as that of the Erie: 28 feet wide at the bottom, 40 feet wide at the water line, 4 feet deep. Almost one third of this project, that began with a decision to adhere to the canal method, consisted of railroad. Going from east to west, there were 82 miles of the

Philadelphia-Columbia Railroad, 172 miles of canal, 36 miles of the portage railway, and finally, 105 miles of canal to Pittsburgh.

With the opening of the mainline in 1834, the confusion and delays that flowed from that hasty decision of 1826 seemed ended; Pennsylvania's interregional transportation problem seemed solved. Yet, only four years after completion of this costly and technically ingenious project, a movement was begun for its replacement. In March, 1838, a Harrisburg convention called urgently for a continuous railroad to meet the competition of the railroad projects of rival states. Surveys authorized by the Legislature during the following years culminated in the report by Charles L. Schlatter, submitted in January, 1842, which showed that it was possible to build a railroad across the state with no inclined planes and with only moderate gradients. Schlatter's report sparked a widespread railroad movement, but the depression of 1842 and the debt incurred in the building of the mainline and the many canal and railroad branches of the state works precluded state action. In these circumstances, there was widespread opposition to a policy that would reduce the state's revenues from the mainline.

Economic recovery, together with the threat posed by the Baltimore and Ohio Railroad's advance toward Pittsburgh, revived the railroad movement, but state construction remained out of the question. However, in contrast to the situation in 1826, when there was no sign of any interest on the part of private investors in construction of the mainline, in the 1840s private construction proved feasible. The Pennsylvania Railroad Company, authorized by an act of April 13, 1846, to build a railroad from Harrisburg to Pittsburgh, received no state aid. Subscriptions by local governments, however—particularly the city of Philadelphia and the county of Allegheny with its cities of Pittsburgh and Allegheny—accounted for over half of the total capital of $12,355,000 in 1856. A continuous railroad from Philadelphia to Pittsburgh was achieved in December, 1852, but the portage railroad of the mainline was still used to cross the mountains and the Philadelphia-Columbia Railroad was still owned and operated by the state. The Pennsylvania's mountain section, which had no inclined planes, was opened in February, 1854.

The curtain rang down on the mainline in August, 1857, when it was sold to the Pennsylvania Railroad for $7.5 million. The Philadelphia-Columbia Railroad was incorporated into the Pennsylvania, and the portage railroad was immediately dismantled. The western canal division of the mainline was abandoned in 1864; part of the eastern division was continued in operation as a canal for local traffic until 1900.[93]

These were the events that followed the 1826 decision to adhere as closely as possible to the canal method. They raise some interesting questions: Was not the decision a serious error? Would it not have been preferable and possible to omit the canal sections phase and to have completed the full railroad in the 1830s instead of the 1850s? Finally, if the decision was an error, how important were its economic consequences for Philadelphia? Two eminent historians have suggested the answers. Firstly, it was a failure: the mainline "never became a strong competitor of the Erie." [94] Secondly, there was an alternative: "Those fourteen million dollars might have been spent to excellent advantage on an all-rail route, which might have done wonders, in the formative years, in diverting the western trade from New York." [95] Let us take the failure first.

Success and failure are of course relative, depending upon the criteria; hence it will be best to measure the record of the mainline against the expectations that led the state government to build the project. The state improvement movement had claimed, (1) that the mainline would earn huge profits; (2) that it would capture the trade of the west; (3) that it would develop the population, agriculture, and industry of the entire state. These will be our criteria. A fourth claim, that the mainline would develop a trade on the scale of that of the Erie, was important in mainline propaganda during 1825 but should not be used as a criterion of the project's success. For there can be no doubt that any Pennsylvania line—railroad, mongrel, or canal— would have failed in such an attempt. The existence of one thousand miles of navigable waterway at the western end of the Erie guaranteed that its success would not be duplicated in the 1830s and 1840s. A failure in this sense was inevitable; it was determined by geography.[96]

By the financial criterion, the failure of the mainline is all too clear. Its original cost was $12.1 million; its total cost through 1857 was $16.5 million; and the excess of revenues over current expenditures through 1857 came to $6.7 million.[97] Mainline advocates, however, did not consider the financial criterion to be the important one. Although they claimed that they expected a great financial success, they were careful to add that a financial failure could still be eminently worth while to the state. We turn therefore to the second criterion: Did the project capture any large part of the trade of the west?

It is true that the mainline, which was soon connected by canal to the Ohio Canal, offered an alternative to the Erie for shippers over a wide area. But it was soon found that the saving in distance gained by going through Pennsylvania rather than through the Greak Lakes and New York was more than offset by the broken character of the transportation line.[98] A Philadelphia businessman, writing in 1859 on the general lack of Philadelphia enterprise, quoted the city's annual import and export figures in the 1830s to prove that the effect of the mainline was imperceptible.

Its capacity for competition in this trade was wholly miscalculated. . . . It was not only more costly, but also to the highest degree, inconvenient to the shipper of bulky products from the West. A milder climate allowed it to be serviceable later and earlier in the season than the Erie; to this incident, and the local trade along its route, its escape from total abandonment is chiefly owing.[99]

In 1839, delays and high tolls on the mainline convinced shippers and millers in western Pennsylvania that it was cheaper to send flour down the Mississippi to New Orleans, then by ship to Philadelphia, than to use the mainline.[100]

The trade statistics confirm these impressions. Through traffic on the mainline, as measured by the traffic over the portage railroad, amounted to an estimated 29,740 tons westbound and

15,439 tons eastbound in 1836 and reached a peak of 83,972 tons in both directions in 1845.[101] This westbound freight was considerable—it was comparable to that of the Erie in those years—and a fairly large number of passengers were carried in both directions. But the west's heavy agricultural goods could not bear the cost of several transfers between railroad and canal. Very little of the estimated 100,000 tons of freight that was shipped annually to New Orleans from the Ohio Valley in the early 1820s was ever diverted to the mainline. In this lay its failure. While New York's Erie Canal–Great Lakes–Ohio canals route took over both the import and the export trade of the northern part of the old west, the southern part continued to use the Ohio and Mississippi rivers for bulk commodities and, to a great extent, the turnpikes for more valuable goods.[102]

However, there is the third criterion: Did the mainline help to develop the economy of the state? Whereas its failure to divert the trade of the west should be measured by the through traffic, which was small, its benefits for the state should be measured by the total traffic, which was much larger.[103] That total traffic reflected the industrial development of Pennsylvania. The mainline received large quantities of coal and raw iron from the Union Canal, from the canals along the north and west branches of the Susquehana, from the valley of the Juniata, and from the huge bituminous coal fields west of Hollidaysburg. A good part of the iron was sent to Pittsburgh for manufacture and then transported east again; and quantities of bituminous coal were shipped eastward to the seaboard cities. In addition, large amounts of salt manufactured in the valleys of the Conemaugh and the Kiskiminetas were shipped to market on the mainline.[104] To the Pittsburgh iron interests, the mainline was a lifesaver. A local improvement would have been adequate to bring the needed supplies to the city, but the undertaking would have been far too costly.[105] The mainline, then, by assembling materials from districts formerly too far apart to cooperate in the manufacturing process, materially aided the industrial development of Pennsylvania.[106]

However, even when this is taken into account, the mainline

was not an unmixed blessing. There can be no doubt that the huge amounts spent on a line that failed to meet the financial and commercial expectations of its sponsors delayed for years the construction of a full railroad across Pennsylvania.[107] The importance of that delay can be gauged from the accomplishments of the trunk railroads in the 1850s. As long as rivers and canals were the primary carriers of inland freight, New Orleans shared dominance with New York in the western trade. It was the trans-Allegheny railroad lines that finally brought the produce of the southern west directly to the eastern seaboard.[108] The Pennsylvania Railroad was able to participate in this development. Like the mainline its principal traffic was determined by the industrial development of Pennsylvania, but in 1860 it transported 176,007 tons from Pittsburgh to Philadelphia, half of which consisted of livestock, flour, and grain.[109] Hence it was only with the completion of the Pennsylvania Railroad in 1852 that the state, "destined by geographic conditions to a land traffic," at last found a method of transportation adapted to its commercial as well as its industrial needs.[110]

A final question remains. Could an earlier railroad have done what the railroad of the 1850s accomplished? If the mainline movement had taken the advice of the railroad party and postponed the entire railroad-canal issue in order to await the outcome of the current railroad experiments in England, construction of a full railroad in Pennsylvania would probably have begun after the definitive and dramatic Rainhill Trials on the Liverpool and Manchester Railroad in October, 1829, the trials that proved the practicability of the locomotive on the general-purpose railroad.[111] In that case, the Pennsylvania Railroad would have been built in the early 1830s. The question then is: Was railroad technology sufficiently advanced by the early 1830s to have produced a railroad that would have avoided the failure of the mainline?

The answer depends in part on the precise reasons for the mainline's failure. Two factors are held responsible: the need for transshipments, and the small capacity and long delays produced by the inclined planes and stationary engines of the portage railroad. A railroad line built in the 1830s would have eliminated the transshipments; it could not have eliminated the inclined planes. If it was the inclined-plane method that caused the failure of the mainline, a railroad project built in the 1830s could not have succeeded.

There is abundant evidence, however, for considering the transshipments to be the principal and probably the decisive disadvantage. As a Philadelphian put it in 1852,

> the chain that was to bind Philadelphia with the west was . . . severed, disjointed, fragmentary. It was an amphibious connection of land and water, consisting of two railways separated by a canal, and of two canals separated by a railway—happily elucidating the defects peculiar to both methods of transit, with the advantages of neither.[112]

So prominent were its disadvantages of delay and expense that it served as a warning against the further building of mongrel lines.[113]

The expense of transshipment was undoubtedly very heavy. Every one of the transportation companies that handled the freight, no matter how small it was, had to have available both cars and boats and had to maintain five sets of depots and agents, located at Philadelphia, Columbia, Hollidaysburg, Johnstown, and Pittsburgh. A businessman with extensive experience in both Pennsylvania and New York testified that the 399 miles of mainline between Philadelphia and Pittsburgh was equivalent in expense to 600 miles of canal, and estimated that maintaining a daily line of freight boats on the mainline "involved a dead loss of about $15,000 when compared with the Erie Canal." [114] An engineer testified that the transshipments caused frequent delays and a high incidence of lost and damaged goods. "The machinery is too complicated for an extensive trade." [115]

The disadvantages were never effectively overcome, despite intensive and ingenious efforts. Canal boats were built in sections capable of being carried in trucks over the portage railway, and cars were built which could be lifted off the wheels and transferred to specially constructed canal boats. But the wear and tear on boats and cars and the amount of dead weight added were serious drawbacks.[116]

The Pennsylvania Mainline

There is room for some difference of opinion as to whether the inclined planes on the portage railway were a serious handicap. One study refers to the portage as "the weak link in the whole system of east and west traffic lanes, in the opinion of all"; [117] and another reports that "to avoid delay and expenses attending the use of the inclines, Philadelphia merchants soon found it to their interest to pay the freight on their goods to New York, and then ship to the West over the Erie Canal." [118] However, these accounts do not distinguish between two causes of delay and expense that were linked to the portage railroad. There was the expense of operation and maintainance and the delay of freight and passengers caused by the ten inclined planes and the use of stationary engines. In addition, there was the expense and delay caused by the need for transshipment at each end of the portage.

Probably the most accurate testimony on this question can be found in the 1846 *Reports* of the Joint Special Committee of the Philadelphia Select and Common Councils, appointed to investigate the desirability of a Philadelphia subscription to the projected Pennsylvania Railroad. The Committee solicited letters from various merchant houses and engineers on the faults of the mainline system. According to the testimony, the serious obstructions and expense were caused, not by the planes and stationary engines, but by the detentions at the points of transshipment.[119] The later accounts of severe criticisms of the use of inclined planes reflected, at least in part, the obsolescence of that method by the 1850s; they are no measure of the role of the inclined planes in the competitive failure of the mainline during the late 1830s.[120]

Still another consideration must be taken into account in gauging the probable effects of each of the choices available to Pennsylvania in the late 1820s. Railroads built in the 1830s required constant experimentation and reconstruction. The early light, flat iron rails were repeatedly replaced with heavier ones until the T rail came into general use; miles of extremely expensive—and rigid—granite roadbeds had to be torn up; roads built on piles to avoid grading had to be abandoned; rails laid on longitudinal beams had to be placed on cross-ties; the feeble, unreliable locomotives of the early days were constantly modified and replaced.[121] These pioneering difficulties, however, would have been well worth while, for the projectors of those early railroads could take immediate advantage of advances in railroad technology. No matter what its immediate success, had a full railroad line been constructed across Pennsylvania in the 1830s, the transition to an efficient line would have been incomparably faster and cheaper.

The evidence is clear. Pennsylvania had a real choice of methods available to her, and a mistake was made. That this mistake had serious results can hardly be doubted. For a decade, Philadelphia suffered the consequences of the realignment of trade brought about by the interregional railroads without receiving any of the benefits. In 1841, the first through railroad over the Appalachians connected Boston to Albany. Boston's merchants, previously forced to sell New England's manufactured goods and to receive its food products through the merchants of New York and Philadelphia, quickly established direct connections with the merchants of the west. Philadelphia lost one of its best markets for breadstuffs and a large part of the trade in New England dry goods and other manufactures that had formerly passed through the city to the west.[122] By the 1850s the opportunity to utilize the advantages of the railroad against New York's canal was lost, for the Erie Railroad and the roads that were to combine into the New York Central were all in operation and connecting lines were being extended to Chicago.[123] Furthermore, there were cumulative effects. One writer refers to the "gloom which pervaded the commercial ranks of society" in Philadelphia when the failure of the mainline to compete with the Erie was followed by severe losses in the China trade: "some of our most astute and enterprising merchants removed to New York." [124]

These consequences provide a vivid contrast with the effects of the Erie Canal decision upon New York. Two of the factors responsible for the difference in outcome have been noted above. First, the success of the Erie Canal produced an extreme sense of urgency in Pennsylvania from which New York had been relatively free during her long period of decision making. Secondly, conditions of geography and settlement in Pennsylvania—and perhaps a more democratic social structure—produced powerful centrifugal forces which made centralized and rational decision making more difficult than in New York. In the Pennsylvania debate, moreover, the first factor tended to exacerbate the second. The felt need for immediate action produced a proposal with such disadvantages that it greatly increased the opposition to any interregional project. These outstanding factors of course do not exhaust the differences. There was evidently nothing in Pennsylvania to correspond to the role that the prospect of federal aid played in New York; and, though it may be added that the canal movement in Pennsylvania produced no personality like that of De Witt Clinton, it is perhaps more to the point to suggest that it offered no comparable political opportunity for a Clinton to exploit.

Was there also a difference between the people of the two states? Were the Pennsylvanians—particularly, of course, the merchant leaders of the improvement movement—innately more cautious, less daring than their New York counterparts? Was that panic reaction and rigid imitation of the New York method partially a result of peculiarly Pennsylvanian traditions and attitudes? A comparison of the improvement literature of the two states seems to support this view. Whereas the agitation for a canal in New York in its final decisive years gives an impression of hard-headed realism—the cost estimates, for example, were remarkably close to the actual cost—Pennsylvania improvement advocacy was characterized by a remarkable lack of realism, both in regard to costs—which turned out to be more than double the estimates—and in regard to the handicap imposed by the mountain barrier. "We have unquestionably the best route for a canal between the eastern and western waters," wrote the majority of the Canal Commissioners to the Governor in November, 1824.[125] It was the "cheapest and best route," the Governor told the Legislature a month later.[126] "The Pennsylvania canal will have immense advantages over the Erie," wrote Mathew

Carey during the debate.[127] No one in the improvement movement, except for the few railroad advocates, ever took up the competitive problem posed by the immense lockage required by any Pennsylvania canal. The Pennsylvania leaders were evidently so affected by the canal mania induced by the Erie's success that the failure of any canal to the west was inconceivable to them. Perhaps the most remarkable statement of their reliance upon the Erie Canal example was produced by Mathew Carey in 1831, when Pennsylvania was already in financial difficulties:

> We must rely wholly on the results of similar undertakings elsewhere, as nearly analogous to ours as possible. . . . Two canals passing through countries of about equal average fertility, inhabited by people of similar habits, manners, industry, zeal, energy, etc., will, in all reasonable probability, afford equal or near equal results.[128]

Unfortunately, these subjective differences are difficult to isolate because the objective situations of the two states were so different. A somewhat clearer indication may be obtained by comparing Pennsylvania's behavior with that of the other major rivals of New York. Only a year after the Pennsylvania decision, a group of Baltimore merchants decided to counter the New York threat by projecting a full railroad to the Ohio. Not until the 1840s was there any comparable willingness in Philadelphia to engage in private construction of an interregional line—with or without state or local aid. Less than a year after the Pennsylvania decision, moreover, the Massachusetts advocates of a line to the west rejected the proposed canal over the Berkshires and suggested a full railroad, but calmly decided, as the Pennsylvanians might have done, to postpone a construction decision until the technical problems of the railroad had been adequately investigated.[129] These comparisons are merely suggestive; but further investigations of such differences of behavior in similar situations may yet produce hypotheses that will begin to answer Jefferson's curious and difficult question concerning "the comparative capability of nations to execute great enterprises."

III. Improvements Without Public Funds: The New Jersey Canals

BY H. JEROME CRANMER

Our city has been swarming with strangers for several weeks past . . . many applicants for laws still throng the lobbies. New York is numerously represented here, and the wall street bees, attracted hither by the savour of New Jersey honey, are buzzing about the State House and looking out sharply for the delicious flowers of good bargains on which to feed and fatten.[1]

To the general rule of government participation in canal construction the state of New Jersey provides a notable exception. In this era of government action in the development of transportation facilities, when New York, Pennsylvania, Maryland, and other states were playing a determinant role in internal improvements, New Jersey's turnpikes, bridges, canals, and railways were exclusively the products of private enterprise. At no time during this period did the government of New Jersey contribute financial aid to the projects within her borders.[2] This choice of free enterprise and steadfast refusal to resort to government intervention has traditionally been ascribed to ideological factors inherent in the hearts of "the strong, self-reliant majority" of Jerseymen. In this view, these rugged individualists "had come through recent experiences with a government which they could not control, and seem to have sensed . . . that a government so organized that it can properly conduct great business enterprises is not safe for democracy." These sturdy pioneers had no doubts as to the ability of private enterprise to accomplish the undertakings, particularly since such doubts "to some extent, cooperated with ideas absorbed from the Communism of Paris." [3] Thus did New Jersey maintain her virtue when other states were bartering theirs away for mere economic advancement.

For both of New Jersey's canals the key decisions against state action or financial aid were made by the Legislature in 1824. Both the Morris and the Delaware and Raritan canals were entrusted to private companies having no access to state credit or state funds. In each case the decision had been preceded by much discussion; it extended in the case of the Delaware and Raritan over a period of more than thirty years.[4] In each case also, the matter was brought to a head by the impending completion of the Erie Canal, its apparent profitability, and its probable impact upon New Jersey agriculture, industry, and trade. The Delaware and Raritan project was for a canal to cross the narrow strip of low-lying land separating those two rivers and to provide an inland connection between the harbors of New York City and Philadelphia. The Morris project, on the other hand, was to proceed from Newark west across the mountainous part of the state to the Delaware River at Easton to connect New York with the coal trade of the Lehigh Valley.

From the time it convened in October, 1824, to the day it adjourned, New Jersey's Forty-ninth Legislature argued canals. Proposals for the accomplishment of the two projects were brought before it almost immediately and were the subject of much discussion and lengthy debate throughout the session. First to appear was a bill to provide for the Morris Canal; this was defeated in the Assembly, receiving support only from the region to be traversed by the proposed canal. The bill for the Delaware and Raritan also languished for lack of support from other than the region it concerned, and it was allowed to lapse in favor of an omnibus bill for the construction of both projects as a single enterprise. This, too, was rejected by the lower house, being opposed by the Raritan party and by representatives of the southern counties. Thus the mutual antipathy of these two competing projects and the opposition of the southern group combined to throttle both canals throughout most of the session. In the closing days of the session, however, new bills were introduced for the Morris and Raritan projects. These in each case called for the accomplishment of the canals by private companies. These bills were adopted by the Legislature and New Jersey was committed to private enterprise.

The purpose of this study is to attempt to discover the peculiar forces or circumstances which enabled—or forced—New Jersey to provide for her two major internal improvements, the Morris and the Delaware and Raritan canals, without resort to state financial aid. In what ways did New Jersey's situation differ from that of the states that followed the general rule of government action?

The Delaware and Raritan

Strange things we are told,
Will come to pass when we are old,
The learned Princeton with renown,
Will then become a seaport town.
If Euclid tires we'll shut the book
And view the fleet in Stony Brook.[5]

The idea of a canal across the state to connect the Delaware and Raritan rivers had been the subject of discussion in New Jersey long before Gallatin's famous report on internal improvements focused national attention upon it. It is reported that in 1676 William Penn and his associates commissioned an investigation to determine whether a canal might be cut from the Delaware to Sandy Hook. In 1774 the suggestion of such a canal had provoked John Pintard, then a student at Princeton, to the verse which appears above. A company had been formed for the purpose in 1796. The project envisioned deepening and improving existing stream beds rather than digging a new channel. Unfortunately, after the company had expended its funds, it was found that whenever the waters were high enough, the streams became too rapid for boats, and particularly rafts, to pass, and the project was abandoned. Eight years later the project was taken up again. This plan, too, called for creating a slack-water navigation by deepening existing streams and building dams to be passed by locks. Although the improvement was expected to cost only $100,000, the company experienced difficulty in obtaining the necessary stock subscriptions, and the project had, once again, to be abandoned.

During the War of 1812, the British blockade of coastwise shipping forced resort to the overland route for traffic between New York and Philadelphia. The consequent inconvenience and expense impressed Jerseymen once again with the need for an inland water communication and added the argument of national defense to the persuasions of economic interest. In 1816 the Legislature named a commission to examine the prospects and to report. The commission rejected previous plans for a slack-water navigation and urged instead construction of a canal to conform with the general plan set forth by Gallatin. This would require a waterway large enough to handle vessels having an eight-foot draft. The cost was estimated at $836,824. No action was taken by the Legislature since many members feared that the cost would be too great for the state's resources. Moreover, a powerful opposition developed from teamsters, stage owners, and innkeepers, who feared that a canal would injure their business.

It was not until 1820 that a bill to incorporate the New Jersey Delaware and Raritan Canal Company was adopted by the Legislature. This charter created a private company having a capital of $800,000, thought to be adequate in the light of the 1816 survey. While the charter was considered "most liberal," there were some major conditions to be met: the canal was to be nine feet deep and fifty-six feet wide; since New Jersey had an agreement with Pennsylvania regarding the use of water from the Delaware River, the company must obtain that state's permission before proceeding; New Jersey reserved the right to purchase the work after fifty years; dividends in excess of fifteen percent were to be shared with the state.

Editorial opinion in the state's newspapers questioned that private funds adequate to the task would be forthcoming since these were depressed years for New Jersey. Hard times, however, could also be used as an argument in the canal's favor.

At a time when labour is unusually low, and the number of poor people out of employ unprecedented, the furthering of this important work ought to claim the attention of all those who are willing to ameliorate the condition of the poor but industrious laborer.[6]

On balance, the conditions and restrictions seem to have heavily outweighed the "most liberal" provisions of the charter as well as the willingness "to ameliorate the condition of the poor." Unable to secure the required agreement with Pennsylvania and unable to find subscribers for the required capital, the company never was formed. In a sense this was fortunate, for in the light of later experience it is clear that four times the capital of $800,000 would not have sufficed to build a canal of the size required by the charter, nor was there sufficient water in the local streams to fill a canal of such heroic dimensions.

Proponents of the Delaware and Raritan had lost a battle but did not count themselves defeated in the campaign. Allowing a year to pass, they approached the Legislature in the fall of 1821 with the hope that the recovery of business of that year together with a more liberal charter would create the conditions necessary for a more favorable response by investors. Once again a bill was passed by the Assembly providing for the establishment of a private company to build the canal. This plan was for a canal having a depth of only four feet, without locks, necessitating a transshipment at each end. Such a work would be far cheaper to build than that of the previous plan, and thus it was thought would have a better chance of attracting the necessary capital. The Legislature felt, however, that a canal of these dimensions would not meet the needs of the traffic, and this project, too, was rejected.

In spite of these reversals, the proponents of the Delaware and Raritan canal continued to urge the project and kept the subject before the public all through the winter and spring of 1822. However, in anticipation of the fall elections and in the expectation that the canal would "excite considerable attention" in the next Legislature, agitation began in earnest late in the summer of that year. Since it was at this same time that "Agrestis" was generating public enthusiasm for the Morris Canal, the newspapers of New Jersey were filled with news dispatches, editorials, letters to the editor, items copied from other papers, all on one subject—canals. The Trenton *True American* expressed great gratification that "the spirit of enlightened improvement is at length roused from the lethargy under which it has until now been oppressed." [7]

Regarding the agency by which the canal was to be accomplished, most papers emphasized the view that the canal should be undertaken by the state rather than private enterprise. Typical of the attitude of the New Jersey press at this time is a statement of the *Fredonian* of New Brunswick.

[The canal] in contemplation from the Delaware to the Raritan ought to be *a State,* and *not a Company,* concern. In the former case, the revenue would be paid in to the Treasury, and lighten the burthens of our people—in the latter, it would go to enrich the monied men of our neighboring states, who would eventually, if not immediately, monopolize the stock, and thus render us forever tributary to them. Is it objected, that the state has no funds? We reply, let it create a stock adequate to the object, redeemable at such time as may be convenient. In this way funds can be instantly provided without the state advancing a dollar. Is it argued that the state cannot manage such concerns as discreetly [sic] and economically as individuals? We answer, New York has proved the contrary as she gets her vast and magnificent work done by contract on terms more favorable than even the lowest estimates.[8]

To stimulate discussion and to excite popular enthusiasm, the New Brunswick *Times* in early August began a series of communications from a correspondent writing under the pseudonym of "Veritas." In his first letter, in order that "the readers of your paper may form some idea of the utility and practical effect of canal navigation," Veritas presented to the editor an article from the Utica *Sentinel* which consisted of a lengthy analysis of the probable revenues for the Erie Canal for that year. From these data he concluded that the income of that canal would greatly exceed the expectations of even its "most sanguine admirers" and that, when completed, the canal would produce for the state of New York revenues sufficient to pay both interest and principal "and then be more than sufficient to defray all the expenses of the state government, and entirely relieve her citizens from state taxes." He pointed also to the developmental effect that the canal was having upon the region "so that in a few years, the whole line of the canal will present an assemblage of villages and a density of population altogether unprecedented."

If the New Jersey canal were constructed and a direct water communication opened between New York and Philadelphia, many of the products brought down the Erie Canal would find their way to the latter place, while great quantities of Delaware lumber, Lehigh coal, flax seed, linseed oil, flour, pork, lard, and butter would move through it to New York. The present huge coasting trade, impressive overland traffic, and immense number of passengers between the two cities would flow through it.

There is, therefore, not the least doubt, that it would immediately yield an interest of at least seven, and perhaps ten to fifteen per cent . . . and before the lapse of half a century the tolls would reimburse both principal and interest, and then be amply sufficient to defray all the expenses of the state government, and relieve the next generation from taxes. . . . It matters not whether a man live in Sussex, or in Cape May, in Cumberland or in Morris, he will participate in the benefits of the New Jersey canal.

Veritas longed for "a Clinton, to rouse the enterprise of our legislature" either to charter a company or to undertake the task itself. If a company were chartered with adequate provisions, it would have no trouble getting the stock subscribed, for several capitalists in New York and in Philadelphia are anxious "to lay hold of it." Veritas would prefer, however, that the state should make and keep it entirely under its own control; there would be no difficulty in obtaining the necessary funds at six or perhaps even five and one-half percent.[9]

Delaware and Raritan partisans lost no time in bringing their project to the attention of the new Legislature. On October 25 a resolution was adopted ordering distribution of copies of the 1816 survey and report to each member. After giving the legislators nearly a week to digest the report, canal proponents secured the appointment of a joint committee "to consider and report on the expediency of the state's constructing a canal to unite the tide waters of the Delaware and Raritan rivers." The *True American* was pleased to state that the subject of the canal was not only before the Legislature but had "engrossed its attention so far." Aware that the Morris project was being agitated, it pointed out that the Raritan would open communication between the two greatest cities of the United States and thus its revenues would not depend upon the "precarious and uncertain quantity of coal to be transported through it to New York" but upon actual traffic "which we know to exist." Moreover, it would pass over a flat, low surface, through light and sandy soil and thus would have no natural obstacles to encounter. Above all, it would confer benefits not only upon New Jersey but upon the entire nation. Clearly the Legislature ought to give its preference to a project so manifestly superior. The Legislature was unable to agree, however, and the project was postponed to the next session.

In anticipation of the fall elections and the convening of the Legislature, agitation for the canal broke forth in late June of the following year and grew in volume and vehemence throughout the fall. The *True American,* for example, in its issue of June 28, inaugurated a series of articles under the heading of "State Concerns," in which it proposed to deal with critical issues confronting the state. The bulk of these columns were devoted to the question of the New Jersey canals. The first communication was that of "Jerseyman" who took note of the continuing proposals of the *National Advocate* that New York State undertake the projects. Jerseyman acknowledged that this would be possible, but, he wrote,

I trust the Legislature of this state will never surrender the privilege of making the canal, and collecting the tolls upon it, and even to a company of our own citizens much less to a company of citizens of another state. That privilege I regard as one of immense importance;

and as it belongs to the state so I think it ought to be exercised for the benefit of its inhabitants at large.

The canal will be highly profitable. Ought the Legislature give the privilege away? Particularly since New Jersey has no "back lands" as so many states have to aid in supporting the government, ought she not undertake the project herself? It was being argued that, if the canal were to be so profitable, why had not private enterprise taken it up under the act of incorporation of 1820—obviously only because private capital was not convinced of the profit prospects. This Jerseyman felt to be "more plausible than sound" in view of the size of the canal required. That plan had called for a canal capable of "sloop navigation," making its cost far greater than those of New York and causing doubt as to there being a sufficiency of water. Clearly it could not be expected that investors would "embark their capital in the undertaking." Given a charter for a canal of dimensions similar to those of New York, which could be built for $250,000, speedily completed, adequately supplied with water, and made to produce a profit within two years, and the stock would be instantly taken up and the work immediately begun. However, Jerseyman hoped this resort would not be adopted.

Every privilege the State possesses ought to be employed to the best advantage for the general benefit. . . . The privilege of making the Canal is worth millions of money to the state; and the Legislature which should attempt to surrender it to individuals, without its full value, would meet with universal and everlasting execration.[10]

The New Brunswick *Times* once again took the lead in supplying information and propaganda materials in the campaign. Early in July "Fulton" entered the lists with a lengthy examination of the profit prospects of the canal whether state or private. The Legislature had for some years been attempting to create a fund for establishing free schools throughout the state. It now amounts to only about $150,000 and is increasing but slowly. Once the net tolls on the canal have paid off interest and principal, they may be turned over each year to the school fund, invested, and allowed to accumulate. Assuming that they will increase five percent each year and that they be invested at "legal interest" (seven percent), in five years they will amount to $325,000, in ten years to $826,000, in twenty-five years to over six million dollars!

[This] would not only be sufficient to establish and maintain free schools in every part of the state, but endow the College of New-Jersey with professorships and scholarships equal to any institution in the world, free the people from all taxation, and furnish the means of making other canals, and railways, and in short, any improvement the state might then be susceptible of, or public convenience require.

Fulton continued with a statement as to the role of government in economic development. "It is the province of great statesmen and provident Legislators, to develope and bring into operation the dormant resources of the community they represent, and to provide not only for the present, but succeeding generations." Unfortunately until recent years New Jersey had confined her views to the present only. However, the subject will unquestionably be considered by the next Legislature, and the question of whether it should be a state or a private concern is a matter of great significance not only for the present but for posterity. Therefore, it is time to examine the question in order that the public mind may be made up on it before the Legislature convenes and that "such unequivocal evidences of the peoples views manifested . . . that they may be enabled to act in unison with the voice of their constituents." [11] The *Fredonian*, on the other hand, saw the principal difficulty in the legislators themselves and their apparent inability to make up their minds as to what course they wished to follow with the canal. "Members are evidently not prepared to make it a State object; nor are they exactly willing that the power of making it a State object should *now* pass out of their hands." Accordingly, continued postponement seemed likely, regardless of the outcome of the commissioners' investigations.[12]

It was recognized that in the Legislature convening in the fall there would be competition between the two major canal projects. "Mentor" in the Trenton *Emporium* urged caution and a careful consideration of each. "Our object should be to begin with the most feasible plan, and [the one] by which the greatest advantages may be gained, in this way the system of Canals may continue popular, and be extended." To illustrate the need for a careful, considered approach, he cites the "Canal mania" in England of some thirty years earlier when "the most ridiculous projects" were attempted and some $100 million spent. Of these projects, some were later abandoned, some were converted into railways, some remain but on an unprofitable basis with stocks from fifty to ninety percent below par, while a few have "amply remunerated their proprietors and extended the most beneficial results to the country." However, in considering this experience,

> Let us bear in mind that in England they have capital to spare, and that rich individuals, expending their money on labourers employed in useless undertakings, is not attended with loss to the public, on the contrary such expenditures are frequently an advantage; but in this country, where capital is scarce, where we annually pay Millions to foreigners for the use of theirs, and where every cent may be employed on works highly advantageous to the community, injudicious expenditures should be guarded against, on principles of political economy.

With regard to the proposed canals, both may be deemed practicable; the people of the state, however, must now "weigh well the supposed superiority" of one plan over another. It was proposed by the canal commissioners in 1816 to build the main canal between the Delaware and the Raritan first and only then to construct "minor canals." "It remains to be seen whether any plans of canals will hereafter be exhibited of sufficient merit to defeat the system proposed by the commissioners and to cause another to be adopted." [13]

The Legislature responded to this agitation by appointing yet another commission to investigate the practicability, probable expense, and anticipated revenues of the proposed canal. The report which this group returned repeated the now familiar arguments for making the canal and for the state's undertaking it. To meet the objection that there was an inadequate supply of water at the summit, the commission accepted the suggestion of the Board for Internal Improvement of the United States Army Engineers that water be brought by means of a "feeder" from higher up the Delaware. This would greatly increase the cost of the project, but, because of smaller dimensions, the total cost would still be about $850,000. The commissioners urged that the state undertake the project by financing it through the sale of bonds.

The Morris Canal

Starting at Jersey City, and running its tortuous course entirely across the State to Easton, on the Delaware, the Morris Canal affords an invaluable means of transit to the highland region of New Jersey, without which its mineral wealth would be entirely undeveloped. . . . Like a broad river, it offers a channel for the internal commerce of the mining and agricultural districts, its benefits ramifying right and left for many miles, every highway and country road forming a tributary, over whose dusty or muddy surface teams loaded with the produce of the farm or the mine, the mill or the forge are hurrying to and fro like a colony of ants.[14]

The idea of a canal across the northern part of New Jersey came to George P. M'Culloch, president of the Morris County Agricultural Society, as a means of ameliorating the difficulties facing farmers and manufacturers of the region in the early 1820s. A near-century of intense cultivation without benefit of fertilizers and the high cost of hauling produce overland by wagon combined to depress agriculture in northern New Jersey. The increasing competition being felt in the New York City market as a result of the opening of sections of the Erie Canal would make it worse. Moreover, the iron industry, which had flourished in the region during the colonial period and particularly during the wars with England, had similarly fallen upon evil days owing to exhaustion of the fuel supply, transportation costs, and increasing foreign competition. A canal seemed to offer a means of redressing the adverse balance for both farmers and iron makers. M'Culloch proposed a canal across the northern part of the state to connect the Delaware River at Easton with the Passaic at tidewater. The summit level was thought to be at an elevation of 185 feet above tidewater and but 115 feet above the Delaware. Thus the total rise and fall to be negotiated would be 300 feet. M'Culloch estimated the total cost at no more than $300,000; gross revenue would be at least $81,000 a year, of which through haul of coal from Easton to New York would contribute the largest amount.

Realizing that public opinion must be marshalled in support of the project, M'Culloch inaugurated in the summer of 1820 a campaign to generate interest in the canal. He began a series of articles in the newspapers of the region pointing out the advantages and urging the construction of the waterway. Signed "Agrestis," these articles were directed primarily toward the farmers of the region and emphasized the probable impact of completion of the Erie Canal. "Within a few months the grand canal of New York will be in full activity; by which grain, salt provisions, butter, spirits, and every other article of value can be brought from the head of Lake Erie to the city, at an expense not greater than carting them from Morristown." Thus the opening of through traffic on the Erie Canal would inevitably depress the level of prices of farm produce to that of the west. A canal would afford cheap transportation for the produce of New Jersey farms and enable them once again effectively to compete in New York City. Moreover, the coal of the Lehigh Valley could be made available to the iron industry and, together with facility of transportation, would introduce new forges, furnaces, and manufactures. "Paterson might then become the Birmingham as well as the Manchester of America." The principal difficulties M'Culloch acknowledged to be the mountainous terrain and the lack of water. As to the first, by the use of locks, tunnels, deep cuts, or inclined planes "the trifling difficulties may be obviated." As to the second, a sufficient supply of water could be made available by damming the outlet of the Great Pond (Lake Hopatcong) which stood at the summit level of the proposed canal.[15]

M'Culloch dealt at length with the agency by which the canal should be accomplished. He objected to entrusting the work to a private company. In the hands of the people the project could be constructed and managed more cheaply and would be of greater benefit to the community than if owned by a company. He cited the charter granted to the Delaware and Raritan Company as an illustration of the reluctance of the Legislature to "cede its authority to individuals," the privileges granted that company were so scanty and restricted as to "frustrate the plan and to prevent subscriptions." Moreover, if the plan is entrusted to a private company,

> Collisions and opposing interests will occur between the state and the company. Unforeseen cases, and exigencies unprovided for, must certainly arise, in which the people will be arrayed against the canal, or the canal against the people. To avoid all these evils, this enterprise should belong to the state; for, however easily a chartered society might be filled up, we ought to be subjected to no influence and bent under no controul, save that of our representatives; nor should our greatest national effort become the sport of speculation or the avenue to a paltry spirit of jobbing.[16]

The enthusiasm generated in the northern counties was intensified during the fall of 1822 when the canal became an issue in the election of that year. The newspapers of the region devoted much space to the procanal discussion throughout the campaign. "Aristides," writing in the Morristown *Palladium of Liberty* noted that the canal would be opposed by the "Raritan influence, which has an adverse interest, in the prosecution of its own canal," and pleaded for the election to the next session of the Legislature of men who would support the Morris Canal, which "must not be huxtered to conciliate influence in distant counties."[17] The adverse interest of the Raritan group had early been expressed in an article appearing in the Rahway *Museum,* which attacked the "Northern Canal" as serving the interests of only one part of the state. The *Museum* listed, among others, the complaints that the Morris Canal presented insurmountable difficulties in construction, that it would provide an intolerable tax burden on the poor, and that the Raritan waterway would better serve the public interest.

Agrestis took up his pen immediately. He pointed out that there are misfortunes greater to a state than loss of power and wealth—loss of sense of patriotism, loss of sense of unity as a state, and the rise of local or sectional feeling. "That our state spirit is extinct, requires no better proof than that, when a measure is proposed which must raise us to our natural level, instead of feeling alive to a common interest, we forthwith sit down to calculate whether some other county will gain more than we." As to the complaint that the Morris Canal would run through only one part of the state,

> I am unable to form any idea respecting a canal which shall run through every part of the state and come home to every citizen, short of converting all New Jersey into a goose pond. . . . If it can be proved that any other [canal] would accommodate a greater number of individuals, have a more extensive influence upon industry, or fix more permanent wealth in this state, I will give up the point.

The Raritan, M'Culloch acknowledged, will be a good speculation for its owners, will be a great convenience to New York and Philadelphia, and "may perhaps transport a few cabbages to a market from some inland situation where they would otherwise have rotted." But its utility cannot be compared to that of the Morris, which will unite coal, iron ore, and water power and thus give rise to manufactures. The Morris project was envisioned as generating a full employment level of economic activity wherein

> every man in this district will find his advantage, every acre of land will be doubled in value, every bushel of grain, every pound of beef will find an instant and high market, every mechanic will find employment, every laborer work, and every manufactory be insufficient to get through its business.[18]

M'Culloch's energetic campaign to popularize the projected canal bore fruit in the fall of 1822 when the Legislature authorized an elaborate survey of the project. The commissioners appointed, of which M'Culloch was one, were given $2000 to meet the expenses of exploring and surveying the route, estimating the cost, and determining the probable volume of traffic and revenue of the canal. A similar canal survey bill for the Delaware and Raritan route was passed by the lower house, but Council postponed consideration of it to the next session of the Legislature. Whereas the sharply competitive nature of these canal projects had early been pointed out, an analysis of the vote in the General Assembly upon the two measures indicates only slight bias for one or the other in particular localities. The lines of cleavage appear to have been drawn primarily upon the question of internal improvements generally and only secondarily upon specific projects. The Morris Canal would pass through the counties of Bergen, Essex, Morris, and Sussex; these northern counties supported the Morris bill unanimously. Opposition came, however, not from the counties through which the Raritan was planned, but from the counties of the southern part of the state. This region feared that state action in internal improvements would result in increased land taxes. Thus the southern counties would be forced to support, but would not benefit from, canals in the north. Most significant local bias appears in the negative votes of Bergen and Essex counties on the Raritan bill. These counties contained the principal commercial towns of the north—Newark, Jersey-City, Elizabeth-town, Paterson—which apparently feared their commercial eclipse by the Raritan towns, New Brunswick and the Amboys, if the southern canal were built.

M'Culloch had not waited for legislative sanction to obtain engineering aid. In October, 1822, James Renwick, Professor of Natural and Experimental Philosophy and Chemistry at Columbia College, and General Joseph G. Swift, former Superintendent of the Military Academy at West Point, joined M'Culloch in exploring the terrain to determine the route which the canal would follow. Both engineers deemed the proposed route suitable for a canal and inclined planes. Renwick submitted a detailed report to the commissioners in which he set forth the most desirable dimensions, the water sources available, the elevation and vertical distance to be negotiated, the cost of construction of the waterway, the practicability of inclined planes, and the economic considerations of the proposed route. His report was eminently favorable on all points.

M'Culloch passed the winter preparing for the surveys to be undertaken in the coming spring and summer. He solicited and obtained the aid of the Canal Board of New York State in detailing one of its most experienced engineers, Benjamin Wright, to assist. He wrote to Secretary of War Calhoun and obtained the services of the Board of Engineers, General Bernard and Colonel Totten. He addressed several requests to De Witt Clinton for a letter in support of the project. Clinton's response is of particular interest for the point of view expressed on the proper agent for accomplishment. "All the canals that have been attempted in the United States, through the intervention of incorporations, have failed . . . principally for want of funds." Should New Jersey attempt the expedient of a private company, either the stock will not be subscribed and the canal never be completed, or, if completed,

> The company will consult its own interests, not the prosperity of the state. The route of the canal will be designated, not with a view to the accommodation of the great manufacturing institutions, but with a view to a cheap, facile, and rapid construction; the tolls may be burdensome, and the superintendence may be vexatious. The cardinal interests of the state may be subordinate to the cupidity of a private association. The capital, if it comes at all, will proceed from abroad, and New Jersey, that has, from the war of the revolution to the present period, evinced a high sense of character and an honorable spirit of independence, will be bound hand and foot by the shackles of a nonresident company.[19]

The Canal Commissioners submitted their report to the Legislature in November, 1823. In preparing their document they had considered the canal from two different points of view: first, as a "project of internal improvement" and, then, as a "measure of finance alone." A favorable vista from either of these positions would justify the canal's construction by the state. Regarding its function in economic development the commissioners asserted that the prosperity which the canal would bring to the northern region was of itself a sufficient motive to "incite its execution as a public concern" regardless of its "fiscal advantages." However, since it will also create a vast revenue, "the enterprise will then appear worthy to call forth the highest efforts of the state."

In describing a country in which industry will flourish, the commissioners sum up the conditions of economic progress. Such a country

> must abound in unwrought materials, it must be plentifully supplied with provisions at the cheapest prices; it must contain water powers to set machinery in motion, and adequate accommodations as to fuel

must exist; finally it must be situated within convenient access to an extensive market. If the canal be made, we are acquainted with no scite so propitious as New Jersey, nor one in which the above described requisites are so eminently combined.

But the canal will not vitalize manufacturing only; it will bring to agriculture the benefits of cheap fertilizers for worn-out soils, of increased local demand by industrial communities, of access to markets of New York and Philadelphia, of increased land values, and

We may here mention as a large increase of freight to the canal, of profit to the farmer, and of good morals to the community, that apples and cider would be exported instead of being converted into ardent spirits. The land carriage renders it at present necessary to have recourse to distilling as the only means of conveying the produce of our orchards to market.

Moreover, the New York canals will soon bring to that city the products of the vast, fertile farmlands of the west, and New Jersey farmers, with high-priced and exhausted lands, will be forced to compete with the "remotest settlers of the new countries, living on a virgin soil, purchased under eight dollars an acre."

The financial resources of the state are adequate to the task. The sale of bonds, banking privileges, lottery rights, and so forth may be resorted to. Most emphatically there is no need for burdening the people with additional taxes. Moreover, the canal will have paid for itself within eight years and will thereafter provide an annual revenue to the State Treasury of $100,000 at the very least. This revenue may then become a circulating fund from which "the School Fund can be increased, taxes paid, roads and canals made, marshes drained, and a system of liberal internal improvement entered upon with spirit and success." The profit prospects of the canal are sure; the report cites the example of fifteen English canals comparing the par value and present market value of their stock and indicating the annual yield on the investment. Present prices are from 20 to 900 percent above par. Annual dividends range from 8 to 119 percent of investment and average 32 percent.

Particular interest attaches to the discussion of the agent by which the canal was to be accomplished. The report asks: "Is the state to call forth its energies to the task, or shall it be consigned to a chartered company?" New Jersey will either follow the example of New York and make public works a state concern or it will establish "the pernicious, the deadly principle of confiding the vital interests of our country to the guardianship of men, whose only view can be pecuniary advantage to themselves." He who controls the source of public prosperity is master of the state and the creation of a private corporation upon which industry will depend risks the hazard of creating "an aristocratic government veiled under the forms of republican administration. . . . Accumulation of money is the sole motive of a chartered company, the extension of industry and happiness is the only legitimate object of a Legislature." Therefore, even if its prospective revenue were insufficient, accomplishment of the measure by the state can be justified on the grounds of public utility. However, since the profits will be sufficient and the benefits to the community great, the state is doubly justified in undertaking the canal.

Despite the fact that the canal would benefit directly only the north, the commissioners had no fear that their fellow citizens of the southern part of the state would regard it "with the contracted views of a sectional jealousy." Shall not all sections benefit from an enriched public treasury? They expressed the hope that an internal improvement for the south would be presented, and in it they pledged the "cheerful cooperation" of the north. In a larger sense, the canal would promote the social connections between "members of the Confederation" by a bond of common interest. It would unite the interests of Pennsylvania, New Jersey, and New York, thus strengthening the national union. For, they pointed out, "commercial intercourse is the most solid basis of political harmony."

The report concluded by answering the argument that the project should be postponed "until the increasing population and resources of the country render the task less arduous at some future time." How is it that in America, where labor is dearest, public works can be accomplished sixty percent cheaper than in England, the parent of internal improvement?

It is because America is a young country, where land is given as a donation, or purchased at a trifling expense; where buildings are seldom disturbed, or are not costly when their destruction is unavoidable; where water can be had without buying up Mill privileges, or erecting steam-engines to pump it from a lower to a higher level. In England every inch of ground must be acquired at an enormous expense; villages must be penetrated, parks traversed, steam engines erected; and circuitous routes adopted. Every rod of land produces a dispute; every mile a law suit; individual interests interfere; the bad passions are roused; and a canal, such as the Erie, would be an impossibility to the English Government with all its riches and power.

New Jersey is now at that optimum moment when population and resources are sufficient and, yet, when improvement and wealth have not erected barriers. Moreover, New York is just now finishing her canal, and "a corps of the most practised engineers in the world, their contractors and laborers, the whole mental and bodily strength of these great works, are as yet unscattered." These may be transferred to operations on Morris. Any delay will disperse them and they will be lost forever. "The words 'Now or Never' may be emphatically applied to our system of internal improvement.[20]

The commissioners' report did not go unchallenged by the Delaware and Raritan party. Probably the most significant and certainly the most trenchant criticism of the Morris scheme came from John Rutherford. Rutherford had long been interested in New Jersey internal improvements. A citizen of Newark, in 1816 he had made a survey for a canal from Newark Bay, across Bergen Neck (Jersey City), to the Hudson. He had long been associated with the Delaware and Raritan project and had been one of the three commissioners appointed in 1816 to make a survey for that route. He was also an associate of men prominent in later movements for the Raritan.

Rutherford's attack first appeared in the Trenton papers. Later it was copied in newspapers of other, particularly southern, communities. It was directed specifically against the report of the commissioners and spared no paragraph of that document. One of the principal objects of Morris, Rutherford began, is to promote New Jersey manufactures by bringing Lehigh coal to the iron works of Sussex and Morris. He questions the value of coal to New Jersey iron masters, pointing out that, whereas it gives great heat and may be used to melt iron "already formed," "no mode has yet been discovered, and is in use, for converting iron ore into iron by means of it." Moreover, since the principal revenue is expected to derive from the haul of coal to New York, it might be well to examine the "conjectures" upon which that expectation is founded. These are that: (1) the use of firewood will be completely discontinued; (2) no coal will be brought from Europe—even as ballast; (3) none from Louisburg, Rhode Island, the Jame River, the Susquehanna, the Schuylkill, or the Lackawaxen will be used; and (4) no coal will be brought from the Lehigh by any other route. Clearly, a revenue depending upon these conditions will never be forthcoming. As to the manufacture of iron, it would be far easier to haul two tons of New Jersey ore to the mouth of the Lehigh than to bring 7.5 tons of coal from there to the mines. Rutherford fears that the mine owners of the Lehigh may combine to raise the price of coal; thus "the profits of the . . . canal may be theirs instead of accruing to those who shall have been at the expense." So far as benefit to agriculture is concerned, the country to be served by the canal does not raise enough for its own needs. Rutherford pointed to the importation of flour from western New York, and asserted that the only market products of Morris County are livestock, "which are driven to market on the leg." Any surplus of bread or grain would be consumed locally by manufacturers.

From an engineering point of view the canal is absolutely impracticable. Many opinions and examples are cited in the report in favor of inclined planes, including that of Fulton and the experience on English canals. However, Fulton had in mind a boat carrying only four tons, not twenty-five as in the report. Also British experience has been limited to descending transport of coal; all other attempts have been abandoned, "with an immense loss." Mr. Renwick's assertions of practicability were so contrary to fact that Rutherford concluded "that he is either desirous to obtain employment for himself as a canal manager or that he is a scientific character let loose from the walls of his study, without practical knowledge or common sense." Rutherford doubted that there would be sufficient water on the summit. Renwick has found it to be sufficient,

but as the Mathematician on another occasion, after traversing the country on foot and in a carriage and astonishing the natives with his scientific apparatus, determined . . . that the summit of the land was only 454 feet above the tide, and now finds it to be about 900 . . . he may also have made a mistake in the quantity of water and admit next year that the amount is not more than half of the estimate of this year.

In noting the present prices of canal stocks in England the commissioners were much too selective. They used only the fifteen *highest priced stocks.* To balance things Rutherford notes the fifteen lowest. Here an investment of £100 would bring but £32 where the report had indicated £100 would bring £609. Rutherford's fifteen are the lowest of fifty-one stocks quoted in "prices current"; the stocks of fifty-two other canals are not even quoted. Presumably they have been abandoned. "Under these circumstances would it be safe to pledge the funds of the state in favor of a project, which will probably cost millions of dollars, and be afterwards abandoned?"

An alternative canal is offered. Rutherford proposes instead of the Morris route, a canal from Easton along the Delaware River to the "Grand Trunk Canal of the United States," the Delaware and Raritan. By such a route the "appalling difficulties of 1644 feet ascending and descending would be reduced to a simple drop of 70 feet from Easton to tidewater." From such a canal Lehigh coal could pass either to Philadelphia via the Delaware River or to New York via the Delaware and Raritan Canal. The latter route might be extended north to Newark, Jersey City, and even Paterson, and south to Camden and the other Delaware towns.

Moreover, Rutherford spied strangers, "persons from other states who would blast our prospects of internal improvements." "The calculating traders of New York" fear that a grand trunk canal through New Jersey would enable Philadelphia and Baltimore to serve as ports for export and import instead of New York. More ominously still, "the subjects of foreign countries" do not wish to see such a canal built since it would act to strengthen the federal union. Rutherford concludes with the prediction that "if the project of Mr. Renwick is carried into effect . . . it will be the most stupendous monument of folly that ever was erected." [21]

Rival canal projectors were not the only opposition which the Morris Canal encountered. The Society for the Establishment of Useful Manufactures feared that the canal would divert Passaic Valley waters from the Society's mill site at Paterson. William Colt, president of the Society, had indicated these fears in a letter to De Witt Clinton wherein he presented a "schedule of manufactories at Paterson" which he feared would be impaired by diversion of water. Clinton's reassurances that adequate water sources existed for the canal which would obviate any recourse to Passaic Valley sources did not satisfy the Society, however, and it continued hostile to the project during the period of construction and the early years of operation.

A joint committee of the Legislature appointed to review the report of the commissioners had no doubt of the practicability of the project. The committee was also convinced that the canal would be a source of wealth and prosperity to the region it traversed. However, as to whether the state should undertake the project, the committee was not prepared to decide. It urged the appointment of a commission to investigate this aspect of the problem. However, no agreement could be reached on this and no action was taken on the project by the Forty-eighth Legislature. The *Palladium of Liberty* reported to its readers: the canal "is considered too important a concern to be thrown into the hands of a private company; and time is required to investigate the resources of the state, preparatory to its being undertaken as a public work." [22]

The commissioners reported, in 1823, and received the thanks of the Legislature for the intelligence, industry, and zeal displayed in the execution of their commission. But that cautious and prudential policy which has hitherto prevented the State from yielding her treasury resources to the blandishments of projectors, charm they ever so wisely, deterred her from making the Morris Canal a state enterprise.²³

Wall Street Bees and Jersey Honey

The important subject of the two Canals, the one from the Delaware to the Raritan, and the other from the Passaic to the Delaware . . . have pressed upon the members from an early day of the session, till near the last one of its final close.²⁴

The Forty-ninth Legislature convened in October, 1824. To judge from the volume of newspaper discussion, the two canals had been the principal issue during the preceding election campaign, and the delegates considered themselves instructed by their constituents on the subject. The Morris project was the first to come to the attention of the legislators. The original intention of its proponents had been to induce the state to undertake the canal, but events in the preceding Legislature had shown this to be out of the question, "through local interests, jealousies, and a most laudable dread of public debt." ²⁵ Because it lacked sufficient provisions for state control of the company the original bill was recommitted. A revised bill reported to the Assembly a week later was the subject of debate for four days and was defeated by a vote of twenty to eighteen, lacking but two votes of the twenty-two necessary for passage. With the exception of commercial Burlington County, the only counties voting for the measure were those through which the canal would pass. Outside of these, only four votes were cast in favor as contrasted with sixteen cast against it.

Meanwhile, though off to a slower start, the Delaware and Raritan project was meeting with better luck than was its rival to the north. Commissioners appointed by the preceding Legislature to examine the prospects of the canal reported favorably and urged its accomplishment by the state. In reviewing the report the Legislature found itself in agreement with the commissioners regarding the practicability and benefits of the canal. Regarding the agent for its accomplishment, however,

It is certainly desirable that so important a link in the internal communication of the country, upon which the interests of other states, as well as our own, will so materially depend, should not be wholly subjected to the arbitrary control of individuals, not even resident, perhaps, in New Jersey, who could break the chain at pleasure, and would thus possess a dangerous power, which might be exercised to favor their own private views, to the serious prejudice of the public.

Also the profits anticipated so confidently provided a strong consideration to making it a state enterprise. However, "although the prospect is flattering," the success of the project is "in some measure problematical," calling for great expense but with uncertain promise of profits. Under these circumstances chartering a private company seemed preferable, particularly since, by reserving a portion of the stock to the state, "most of the benefit expected may be realized to the state, without assuming the great responsibilities which must be incurred."[26] Accordingly a bill was presented to incorporate a company to make the Delaware and Raritan Canal. The bill reserved to the state the right to subscribe to one eighth of the stock within one year and to take the canal after fifty years. No action was taken on the bill however; other and more momentous, or at least grandiose, plans were under consideration.

The proposal to charter the New Jersey State Canal Company, or "the Mammoth Plan," as it came to be called, was the creation of the Morris proponents. Feeling that the failure of the original Morris bill had been due to "an unfortunate collision with the advocates of the Raritan canal" and implying that the reluctance of the Legislature to undertake the Raritan was due, in part at least, to hostility from the north, they sought a means to unite the two forces in support of one bill. Three major objectives were set forth for the plan. First, it would obtain for the people one third of the ownership of the two canals, would secure to them a control in their management, and would provide an opportunity for the state to make them an exclusively public property in ninety-nine years. Secondly, "by executing, at the same time, both the Morris and Raritan canals, the present plan terminates all unfounded jealousy between the friends of each respective enterprize, and extends almost over the whole State, the advantages of inland navigation." Thirdly, it concentrates in New Jersey several millions of capital from abroad and accomplishes "at foreign expense, an extensive system of improvement." It will bring into the state treasury a net income of $20,000 annually, which will almost certainly increase to $100,000 as soon as the canals are opened. How were these eminently desirable goals to be achieved? Essentially the state would grant liberal banking privileges to a private canal company. Thus the state would exchange the charter of a huge bank for construction of, and control over, the two canals.[27]

As it was brought before the Legislature, the Mammoth Plan called for the creation of a company having $6 million of capital, $4 million to be subscribed by private individuals and the remainder reserved for the state. However, the state's portion was not to be paid in cash but would consist of a loan of $2 million of "certificates" (state bonds). These would bear five percent interest which would be paid by the company. The company would also pay to the state each year one percent on the value of the certificates. After ten years the state would have the option of accepting cash or stock in exchange for the certificates thus loaned. Supervision of the company was to be centered in a board of ten commissioners, six to be appointed by the state, and four by the company. Management of the company would be in the hands of two boards of directors: one for the canal and the other for the bank. The state was to name three fifths of the directors of the canal board but none of the directors of the banking portion of the company. On pain of forfeiture of the charter and of the stock subscription the company was to expend at least $300,000 each year on construction of the two canals. The company was granted full banking privileges for all capital not expended on construction, and, although tolls were to be fixed and canal operations strictly supervised by the board of commissioners, the company would have a completely free hand in the use of its "surplus capital" in banking operations. No supervision or control of the bank by the state or by the proposed board of commissioners was authorized; private stockholders alone would control the banking operations of the company. These operations would include stock brokerage, exchange brokerage, loan, insurance, and pawnbrokers' offices, as well as the more normal banking activities. A single banking office was to be established at Jersey City.[28] It is clear that the plan anticipated granting tremendous banking rights in return for assurance that the canals would be constructed and provided for state control over canal operations but no control over the company as a bank.

The Legislative Council accepted the plan and resolved to draft a bill "embracing the principles contained therein," and named representatives to a joint committee to write the bill. The lower house, however, rejected the report and, when informed of the intentions of the Council, took the apparently unprecedented step of refusing to appoint members to the joint committee. Fortunately the "yeas and nays" were required, thus affording an opportunity to determine the areas of support and opposition. On the question of agreeing to the resolution to appoint a committee to meet with that of Council, only twelve votes were cast in favor, twenty-seven members were opposed. Support for the measure was confined to Bergen, Essex, Morris, and Sussex counties and appears to have come from two diverse interests. One was the canal group of the northwest, Morris and Sussex counties, which tolerated banking operations as incidental to the principal purpose of canalling. The other group, Bergen and Essex counties, was interested in obtaining banking privileges and viewed the canals as a means of extracting those rights from a reluctant Legislature. It looked to the creation of a $6-million banking company which incidentally would construct canals. This latter group was supported, perhaps dominated, by New York banking and speculative operators. Opposition to the measure also centered around two groups. One, the backers of the Delaware and Raritan canal, not wishing to see their project taken over by northern or out-of-state bankers, and fearing that state support for Morris would prejudice the chances of their canal, joined with the southern agricultural group to defeat the measure. The *Centinel of Freedom* reported that

> the great and splendid and, to our little minds, overpowering scheme of making the contemplated improvements in our State by means of

a six million bank . . . has been disposed of rather unexpectedly, and in a very quiet, snug, and suitable manner. The honorable Council, with more *capacious* minds than their brethren below stairs, seemed fairly to have embraced the notion that the aforesaid nondescript establishment benevolently offered facilities for our improvement and prosperity, which it would be nothing short of treason to the State to reject—and they therefore [resolved] to bring in a bill to carry into effect the benign intentions of our friends from Wall Street, New York. The House, however, to the utter astonishment and confusion of the *higher* power, and with singular and invincible indifference to the rules of common courtesy and coordinate deference, declined compliance with the request of Council—and thus the infant Hercules was strangled in his birth!

Having recovered from the shock which the crush of this machine occasioned, we are now plodding on in our old two-penny calculating way, to endeavor to get our canals made by dint of labor and a *little* money.[29]

With the defeat of the Mammoth Plan the separate measures for creating private canal companies were brought once again to the attention of the Legislature. On December 16 the Morris bill which had been lying "dormant on the files of the House," was called up and sent back to committee. There it was amended to include banking privileges to the extent of $1 million. Whereas the original act had proposed a board of directors to be composed of two members to be appointed by the state from each county through which the canal would pass, the charter was amended to allow the stockholders to elect their own board. Thus control of the company was removed from the hands of the state and placed in those of the banking interests. The bill was reported out of committee the next day and the House devoted the following week to consideration of it. The measure was passed by the Assembly on December 22 and was sent to Council. Here, after further minor amendments, it was approved and returned to the House for final adoption, which it received on December 29.

The bill to incorporate a company to make the Delaware and Raritan canal had lain dormant during the period of discussion on the Mammoth Plan. Apparently the promoters of this bill now increased the attractiveness of their appeal by offering a bonus of $50,000 to the state in return for a charter. Anticipating the failure of the measure, a second appeal for a charter had resulted in the introduction of a bill to incorporate the New Jersey Canal Company; it, too, was concerned only with the Raritan route and was considered rival to the first. This company, however, offered a bonus of $60,000 to the state. The Assembly passed the latter plan on December 22 and forwarded it to Council. That body, to the suprise of most observers, postponed the measure indefinitely. On December 28, however, three days before the Legislature was to adjourn, a new bill was introduced into the Assembly. This measure was somewhat more generous in its treatment of the state treasury for the bonus was up to $100,000. With the opposition of the Morris group now removed, this bill swept through both houses and received final approval in a special evening session of Council on December 30.

An analysis of the votes in the General Assembly on final passage of the two canal projects indicates surprising unanimity in view of the prior hostility manifest in the voting record. The distribution of votes was almost identical; no county changed its vote on the two bills. Bergen, Essex, Morris, Middlesex, Burlington, and Salem were unanimous in support of both projects; Hunterdon and Somerset were unanimous in opposition to both. Of the thirty-one members voting for the Delaware and Raritan at least twenty-four had voted for the Morris Canal. It would appear from this, and from the dispatch with which both bills moved through the Legislature, that a reconciliation had been reached between the two canal parties. Moreover, the substitution of banking privileges for state enterprise in the case of the Morris Canal and the tender of $100,000 to the state in the case of the Delaware and Raritan, had persuaded the southern agricultural countries to support internal improvements in New Jersey.

Comment on the passage of Delaware and Raritan charter largely concerned the wisdom of entrusting the task to a private company. The *Centinel of Freedom* was convinced that the bill passed Council only because the bonus had been raised to $100,000. It also felt that promoters of Morris had obstructed passage, allowing it to progress only "as their own was suffered to go on." The *New Jersey Journal* also took note of what it termed "a respectable but small minority" which had opposed the bill on the grounds that the state should undertake the canal and a "most powerful opposition" to both canals generated "out of doors" by agents of the Delaware and Hudson. Always an advocate of state enterprise, the *Centinel of Freedom* commented somewhat bitterly that, although the state had attempted to protect its interests, nevertheless,

the managers in this business have completely put the screws upon the state; and when we take into consideration the admirable tact, address, and undermining talent of our lobby in the halls of the New Jersey Legislature, we shall not be surprised at the issue.[30]

The charter of the Morris Canal represented an attempt by the Legislature to grant as liberal a charter as possible in order to encourage private venture capital consistent with retaining as great control as possible over the company in its canal operations. It is clear that while the Legislature regarded the benefits to be derived from the canal as eminently desirable, it was, nevertheless, not willing to undertake the project itself. Nor was it willing to resort to a mixed corporation. On the other hand, it had been demonstrated that the high risk element and the relatively low return of canal operations would not prove sufficiently encouraging to private investment funds to entice them away from the more lucrative land speculation, insurance investments, or banking operations. It was under these considerations that the Morris charter was framed. The subscribers to the stock were to be incorporated as the Morris Canal and Banking Company; subscriptions were to be in amounts of $100 to the sum of $1 million for canal purposes. If necessary for canal construction, this could be increased by an additional $500,000. Six commissioners were appointed by the charter to receive subscriptions and organize the company.

Article xiv provided for the right of the company to banking activities.

> For the encouragement of so great an undertaking as the erection of said canal, and in some measure to induce capitalists and others to subscribe to the same, it shall be lawful for the said company to issue stock for the purpose of banking operations, but under the following conditions, limitations, and restrictions.

Each expenditure of $200,000 upon the construction of the canal entitled the company to call upon the stockholders for subscription to an equal amount of bank stock. The maximum of bank stock which could be sold was $1 million. Thus only after actual expenditures upon construction had been made could the company make use of its banking privileges. Moreover, to continue these privileges the company must continue to make expenditures upon construction until the canal be completed. The grant of banking rights was limited to a period of thirty-one years.

The company was granted other valuable rights and privileges. It was exempt from all taxes other than the customary state tax on bank capital. It was granted the right of eminent domain. Any land, water, and streams useful for the canal or materials necessary for construction on adjacent lands might be appropriated. The company could establish its own tolls subject to a maximum of three cents per ton-mile. The company was granted a monopoly; no other canal might be constructed within ten miles. The state reserved the right to purchase the canal at the end of ninety-nine years at "fair value." Should the state not exercise this option, at the end of another fifty years the canal would automatically revert to the state.[31]

Thus the Morris Canal and Banking Company was the offspring of rather mixed parentage provided by the banking party of New York and the canal party of northern Jersey. It was evident in the company's early years that the speculative genes of the banking parent were dominant. Subscription books were opened at Jersey City in 1825 and, "amid scenes of the greatest excitement," the $1-million capitalization was heavily oversubscribed. Reports of the amount of oversubscription ranged from ten to fourteen to one. The $100 par shares soon claimed a premium of twenty percent. The dominance of the banking influence was evident in the board of directors named at the first meeting of the stockholders. Of the fifteen directors, at least nine were from New York; only four are known to have been Jersey residents. Within two years, four of the nine were indicted for fraud in connection with their dealings in Morris stock. Moreover, although the company's single office was located in Jersey City, its main business connections were across the Hudson in New York. The emphasis upon speculative banking rather than canalling was particularly marked during the middle 1830s. Under the presidencies of Louis McLane and E. R. Biddle, nephew of Nicholas Biddle, close relations were established with the Second Bank of the United States, then operating under Pennsylvania charter. The most notorious of the company's operations concerned the sale of internal improvement bonds for the states of Indiana and Michigan. So complete was the emphasis upon banking that in 1836 the canal was leased to the Little Schuylkill and Susquehanna Railroad. This, according to McLane would produce a return of four and one-half percent per share but, more importantly, would enable the officers to give closer attention to the other affairs of the company.

Construction was begun in 1825 but was delayed by lack of funds since actual costs far exceeded estimates. By the time it was completed from Newark to Phillipsburg the canal, which had been expected to cost $800,000, had cost over $2 million. Loans and additional stock issues provided the required funds. By 1837 the canal had been extended to Jersey City, total capitalization was over $4 million, and the funded debt exceeded $2 million. The company failed in 1841, and the canal was sold in 1844 to a new group interested only in transportation. The canal was enlarged and during the decade of the 1860s proved profitable, earning, in 1864, nearly seven percent on the cost of the canal. As had been anticipated, its principal traffic was in eastbound shipments of coal to meet the needs of New Jersey's growing industry and in westbound shipments of iron ore from Jersey mines to Lehigh Valley iron and steel mills. In 1861 the

Morris and Essex Railroad pushing west across the state reached Phillipsburg; seven years later it became part of the Lackawanna, providing a through rail link from Scranton to tidewater. In the face of this competition the canal was doomed, and, while it continued to struggle for another fifty years, it passed out of existence in 1929.

The charter of the Delaware and Raritan Canal Company was designed to shift the burden of financial risk onto the shoulders of private enterprise while reserving as much as possible a major share in the revenues for the state. Capitalization was $800,000 and might be increased to $1 million if needed; one fourth of the stock was to be reserved for the state. Management was vested in a board of seventeen managers, a majority of whom must be residents of New Jersey. As with prior companies chartered for the Delaware and Raritan route, regressive voting rights were provided in a measure to reduce the danger of "foreign" control. The company was to be free from all state, county, and local taxes on its canal properties and operations; it was obliged, however, to pay the bonus of $100,000 to the state. Eminent domain, use of materials locally available, and methods for the condemnation of property were provided. Tolls set by the company were not to exceed two cents per ton per mile on coal and four cents per ton per mile on merchandise; the Legislature reserved the right to appoint commissioners further to supervise and establish maximum rates of toll. No other canal or railway could be constructed within ten miles of any point on the canal or feeder without the company's permission. The state reserved to itself the right to take possession of the canal after fifty years and upon payment of the "first cost" of the canal.[32]

In order to provide sufficient water at the summit level the charter of the Delaware and Raritan called for a "feeder" to tap the Delaware River and conduct the water to the main line of the canal. The company must obtain Pennsylvania's permission for such diversion before it could proceed.

This permission was obtained, but, because of the opposition of Philadelphia interests, who feared a diversion of Delaware River trade to New York via the feeder, the grant was so hedged about with restrictions as to be almost useless to the company. Although they felt that no canal could be built under these restrictions, the managers resolved to proceed with the organization of the company hoping that a more liberal grant could be obtained the next year. Accordingly subscription books were opened at Trenton in May, 1825. The response there was pitifully small, owing probably to the influence of Philadelphia; just four shares were taken. When the books were opened in New Brunswick, however, the response was overwhelming; forty-eight subscribers undertook to buy a total of 77,480 shares. Each of twelve persons wanted to take all 6000 shares. The bulk of the stock was reported to have been taken by New York and Boston capitalists. The company was organized; the $100,000 bonus was paid to the state; and the engineers were directed to proceed with a definitive survey.

Despite intensive lobbying, political pressure, and public agitation, hopes for a more liberal grant from Pennsylvania were disappointed. An amended bill was introduced at Harrisburg, thanks to the support of the upper Delaware Valley, but the hostility of Philadelphia could not be overcome, and the final bill represented no significant improvement. A report from the company's engineers that the original cost estimates had been far too low and that a canal of the dimensions required would cost over $3 million seems to have supplied the final disappointment which dashed the hopes of the New York and Boston stockholders. Though the directors from New Brunswick, Princeton, and Trenton seem to have been eager to continue, the out-of-state directors urged abandonment. The latter view prevailed and the company petitioned the state for return of the bonus. After lengthy discussion and much debate this was approved by the Legislature and in 1828 repayment was completed and the company was dissolved.

With the end of the speculative boom after 1825 and the ensuing abandonment of the Delaware and Raritan by private enterprise, attention in New Jersey turned once again to the possibility of state action. The precedent for state action provided by

New York State, especially its role in the Erie Canal, was powerfully reinforced in 1827 by Pennsylvania's adoption of an elaborate program of internal improvements. Moreover, the threat of the proposed Delaware Division Canal was not lost upon Jerseymen, nor was the fact that that portion of the Pennsylvania system would need water from the Delaware River. That the roles had been reversed and that Pennsylvania was now the suppliant for water rights clearly implied that a far more favorable agreement might be concluded. Accordingly, hopes for the long-delayed accomplishment of the Delaware and Raritan Canal ran high.

The 1827 campaign to induce the state to undertake the project got under way in August in anticipation of the fall elections and the session of the Legislature. Noting that the way for state participation in Pennsylvania had been paved by the internal improvements convention in Harrisburg, proponents of state action urged a similar meeting. The New Jersey Convention for Internal Improvements met at Princeton in September. Committees were formed to draft resolutions expressing the sense of the meeting on internal improvements, to prepare memorials to the Legislature on the subject, and to frame an address to the people of the state. All concluded that the state had been too neglectful of her interests in the development of transportation facilities, that the best interests of the nation and the state would be served by the canal, that in the present condition of the money market the necessary funds could be borrowed at a mere five percent, that the expenditure of these funds would be of immense benefit to the citizens of the state "at this depressed time," and so forth. These statements received wide attention and discussion during the election campaign that fall.

Upon convening, one of the first acts of the Fifty-second Legislature was the appointment of a committee to investigate the expediency of the state's undertaking the Delaware and Raritan Canal. This group reported to the Assembly their conclusions that the canal should be built and by the state. The canal was necessary to complete the chain of many thousands of miles of navigation—a chain interrupted only by the short space across New Jersey. New Jersey ought do its part, particularly since it would be very much in its own interest for it to do so.

We invariably find, that manufactures, trade, and agriculture, flourish most in those states where the government has led the way in public improvement—where the spirit of those in authority has drawn forth the resources of the state to contribute to its internal improvement, and given the lead to the exertions and enterprise of its inhabitants. The State of New Jersey ought to make the canal.[33]

The committee returned a bill to that effect. This bill was entitled "an act to provide for the improvement of the internal navigation of this State" and in effect provided for the establishment of a canal authority to undertake the work for the state. The Governor was to appoint five commissioners who would be responsible for hiring engineers to survey the route, for engaging contractors to undertake construction, for hiring employees to operate the canal once completed, and for establishing rates of toll and making collections. Funds for construction were to be provided through the sale of $1 million of five-percent bonds on the property of the canal and backed by the credit of the state. All tolls over and above maintenance and operating costs were to be dedicated to payment of interest and principal, as were the proceeds of the bank stock tax. Thus the canal was to be built by the state without recourse to taxation other than the bank stock tax. The press of business before the Legislature necessitated an adjourned session, and in mid-November the first sitting was terminated to reconvene in mid-January.

The interval between sittings of the Fifty-second Legislature provided an opportunity for public discussion of the principal matter before it, the Delaware and Raritan Canal. The public journals were full of letters and editorials, announcements and reports of meetings grappling with the question: Should the state undertake to build the canal? The case for the opposition was developed by "Jerseyman" in a series of letters appearing in the *Federalist*. Naturally these letters were widely reprinted and were the subject of much discussion and many replies. In urging a calm and dispassionate investigation of so important a subject,

The New Jersey Canals

Jerseyman admitted that he was expounding an heretical view at variance with popular opinion. Such was the state of feeling to which the citizens of Trenton had allowed themselves to be wrought, that any expression of doubt as to the advantages or hint at the possible inexpediency of the proposed canal was deemed equivalent to an "admission of idiocy or a declaration of treason."

The Public mind seems to require some subject of excitement; some topic upon which extravagance may exhaust itself. And the hobby which at present engrosses attention is the formation of canals. The successful result of one of these projects, under circumstances favorable beyond example, has set the nation wild, without any regard for the analogy of the cases, or the obstacles which nature has placed in their way. More of these schemes have been [projected] than we or remotest posterity will ever see accomplished.

It was to be hoped that New Jersey would not "rush blindfold" into such a project but would examine the factual data on the subject, not the estimates of the "fertile imaginations of engineers drawing upon the future wealth and resources of the country, quickened by the prospect of profitable employment, but calculations founded upon what has been done before." Particular emphasis should be placed on the experience of New York with the Erie Canal since the advocates of the New Jersey canal rely upon that example as their strongest argument. To do so indicates "a lamentable ignorance of the great and irreconcileable differences between the two undertakings." Here Jerseyman suggested a basis for distinction between canals. On the one hand are those built primarily for the purpose of economic development of a region; on the other are canals built with the object of exploiting an opportunity for profit. Income from tolls was but the smallest part of the recommendation of the Erie. The primary advantage of that canal was that it opened a "vast and fertile country" to settlement and commerce, which otherwise would have remained undeveloped "for centuries to come." The canal also opened for development the country bordering on the great lakes, whose two thousand miles of shoreline were surrounded by upwards of fifty million acres of land. Obviously a canal opening up such a region would give value to lands previously valueless, would create towns along its banks "as though by enchantment," and would make New York City the principal commercial center of the United States. Contrast this with New Jersey. No point on the proposed canal is at present more than twelve miles from navigable water. There simply will be no development effects from the provision of transportation since cheap transportation is already available.[34]

When the Legislature reconvened in January it was met by a veritable flood of petitions for and against the bill. The bill was defeated in Council, however, by a vote of seven to seven. There being no majority, the bill was not passed.

In addition to the Delaware and Raritan Canal, there were applications for charters for railroad companies brought before the Fifty-third Legislature. All told there appear to have been four different plans for railroads to connect the Delaware River with the port of New York. Most important of these was the petition for the proposed Camden and Amboy—a project to connect the Delaware opposite Philadelphia with the Raritan and New York. This measure was rejected by the Assembly and an examination of the votes of individual members on the canal plan and this railroad project indicates significant mutual antipathy between them. Of the twenty-two members who voted for the bill for the state to build the canal only six voted for the railroad, while of the nineteen who voted against the canal fourteen favored the railroad. Thus a significant source of the opposition to state action on the canal in this session was the rivalry of the railroad party. Both projects were brought before the Legislature the following year with virtually identical results. The canal project passed the Assembly but was rejected in the Council, lacking one vote of passage. The Camden and Amboy bill was defeated in the Assembly.

Thus the movement for New Jersey to construct the Delaware and Raritan Canal was defeated by the narrowest of margins in 1827 and again in 1828. This was the closest the state ever came to direct action in internal improvements during the period. Failure on each occasion appears to have been due to a combination of forces by the southern agricultural counties and the supporters of the Camden and Amboy railroad project. Local jealousies and conflicting economic interests, then, were the main factors responsible for the rejection of state action in these two years.

Undaunted by their repeated defeats and encouraged by the very narrow margins of their recent ones, the proponents of the Delaware and Raritan resolved to offer a somewhat modified plan to the next Legislature. This proposal sought, in the words of the *Emporium,* "to avoid the rocks on either hand upon which this great project has so often been shipwrecked." Every prior scheme had relied on one of two means: (1) incorporating a private company or (2) undertaking the work on state account. The principal objections to the first were that it placed too much power in the hands of private individuals and that the "immense profits certain to result" legitimately belonged to people and ought not be given away. On the other hand the state had been objected to as entrepreneur because it was without resources other than credit and $1 million was thought too much to borrow. The plan being put forward currently was a compromise measure which sought to divide the burden between state and private enterprise—a resort to the Virginia plan of "mixed enterprise." Under the plan the Board of New Jersey Canal Commissioners was to be incorporated at $1 million, one half to be subscribed by the state, the remainder by the federal government and private capitalists. The state was to borrow the funds necessary for its subscription. In this manner it was hoped that the state could retain control and obtain one half of the profits while being relieved of one half of the expense and debt burden, thus allaying the fears of the "most timid." [35]

This plan, however, did nothing to mitigate the hostility of the Camden and Amboy partisans. Fearing that once the state interested itself in one transportation facility it could never be persuaded to grant a charter for a competing one, the railroad party remained adamant in its opposition to the canal.

So evenly were the forces divided that either, with the support of the conservative southern counties, appeared able to defeat the other and a deadlock seemed once again in prospect. The Society for Internal Improvements carried on an active campaign of pressure and propaganda against the railroad, emphasizing that it would become a monopoly and would destroy the business of taverns and stage companies, while the canal would be a public road and have little effect on the transport of passengers. The railroad party, on the other hand, pointed to experience in Great Britain to show that railroads will entirely supersede canals. Indeed it was fortunate that the state had not previously engaged in so obsolescent an enterprise since that probably would have stood in the way of a railroad. Both parties retained powerful lobbies at the capital and so bitter was the feeling that the leaders felt it necessary to be armed when walking the streets of Trenton.

According to the oft-repeated story, the legislative log-jam was broken by a chance meeting between Robert F. Stockton of the canal party and Robert and John Cox Stevens of the Camden and Amboy in the foyer of the Park Theater in New York. This encounter led to conversations which ultimately produced an agreement between the two forces. Some writers doubt the authenticity of this story, largely on the ground that no overt agreement was achieved until more than a year later. However, while the canal bill was still under consideration by the Assembly, it was abruptly withdrawn, returned to committee, and on the same day a new bill was reported out. The new bill abandoned state action and called for the incorporation of the Delaware and Raritan Canal Company. From this moment on, the two bills moved through the halls of the Legislature with suspicious dispatch and uniformity. On January 28, both were called up and passed by the Assembly, the canal by 22 to 17, the railroad by 28 to 15. Of the twenty-two members who voted for the Delaware and Raritan, eighteen voted for the Camden and Amboy. When the scene shifted to the upper house a similar drama of expedition and regularity was played. Both bills were received on January 28 and committed. Both were reported out of committee the next day, both without amendment. Both were considered for one day, and both, after a motion to postpone had been defeated, were passed on February 4, 1830,—the Delaware and Raritan by 8 to 6, the Camden and Amboy by 10 to 4. Of the eight Councilmen who voted for the Delaware and Raritan all voted for the Camden and Amboy.

The charters granted the two companies mirrored the uniformity of their passage through the Legislature. The only differences were the result of the different transportation media to be employed. Both of the companies were capitalized at $1 million with the provision that this might be increased to $1.5 million, if necessary. In both cases, the state reserved the right to subscribe for one fourth of the original stock: in the case of the Camden and Amboy the reservation would expire after one year, in the case of the Delaware and Raritan, in two. The state also reserved the right to purchase the properties "at a fair value" thirty years after their completion. Maximum tolls were established for both: the Camden and Amboy could charge no more than ten cents per mile for passengers and no more than eight cents per ton per mile on freight; the Delaware and Raritan was held to five cents per mile on passengers and four cents per ton-mile on freight. Both were granted monopolies: no railroad could be built within five miles of the Camden and Amboy without the permission of the company, and no canal could be built within five miles of the Delaware and Raritan or its feeder. As a *quid pro quo* for this monopoly grant, and in lieu of all taxes, the companies were to pay transit duties to the state: the railroad was to pay fifteen cents for each passenger and ten cents for each ton of freight, the canal would pay eight cents for each passenger and ton of high-grade freight and two cents per ton of low-grade freight, such as coal. These charters established a precedent in New Jersey which the state found it difficult to reverse in later years. The state had made a bargain with the companies granting them valuable privileges in return for revenue. It was entirely possible that the state would desire to increase those revenues by the sale of even greater privileges. Mindful of this the companies might well seek to tempt the Legislature in the future.

Neither the canal nor the railroad was entirely satisfied with its original charter. Well aware of the competition which the railroad would offer, the canal company sought permission to build a railroad to parallel the canal. The Camden and Amboy could not allow this, and so the Legislature refused to make such a grant. The railroad, on the other hand, aware of the many petitions before the Legislature for competing lines across the state, sought to increase the monopoly provision of its charter. In return for a gift to the state of one thousand shares of its stock, the Legislature agreed that no other railroad should be authorized on the pain of forfeiture of the shares. This amendment was accomplished with the support of the Delaware and Raritan group, for by this time a permanent agreement had been reached to unite the two companies into one "mammoth monopoly." Permission to accomplish this union was granted by the Legislature in February, 1831. The right of the state to subscribe to one fourth of the stock was not acted upon. Instead, upon the gift of another one thousand shares in the combined company and upon the assurance of a minimum of $30,000 annual income to the state from transit duties and dividends, the Legislature granted an absolute monopoly of New York–Philadelphia rail transportation across New Jersey to the Joint Companies—as they came to be called.

The canal was opened in the spring of 1834. Although tonnage and revenues increased rapidly, over the first ten years, it earned less than one percent on its cost. By 1850 it was earning three percent and by 1860 over ten percent. As with the Morris Canal, the Civil War brought greatly increased traffic; the Delaware and Raritan earned twenty percent in 1865 and nearly twenty-four percent in 1866. Unlike the Morris, however, the volume of traffic remained high after the war, and operating profits continued above ten percent through 1871. In that year, the Pennsylvania Railroad, seeking access to the Hudson, leased the properties of the Joint Companies. The canal remained in operation into the twentieth century. However, by 1900 it had ceased to be profitable, and in 1933 the Pennsylvania surrendered the property to the state.

The New Jersey Canals

Exploitation vs. Development

Canals, when considered by a comparison of their cost with the revenue derived from them, may be divided into two classes. 1st: Those which are made with a view to the general interest of the country, the revenue being a secondary object; and 2nd: Those in which the revenue is the principal object. The first can only be undertaken at the charge of the public treasury; the other may be either the work of the nation, of particular states, or of private associations.[36]

In considering the basis for New Jersey's decision regarding state aid to the Morris and the Delaware and Raritan canals, it is useful to recall the distinction that was repeatedly drawn between kinds of canals by writers at the time. The distinction lay in the economic function to be performed by the canal. In this view a canal would be worth while if it provided transportation facilities to an otherwise isolated region and by this means stimulated the economic life of the region. Such a canal may be characterized as a "developmental" project—that is, a project the principal function or purpose of which was the economic development of the region traversed or reached by the canal. On the other hand, a canal would be justified where the geographical and economic circumstances provided a profit opportunity to be seized. This latter may be considered an "exploitative" canal—that is, a project the function of which was to take advantage of an existing profit opportunity. This distinction was drawn repeatedly during the debate in New Jersey regarding the Morris and the Delaware and Raritan canals.

The distinction between developmental and exploitative canals is of significance for the light it casts upon the attitude toward the role of government. In the case of a waterway the purpose of which was to open up a new region to trade and commerce, it was clear that profits would have to await the anticipated development. This might proceed quite slowly and it be many years before the canal would produce revenue sufficient to meet current operating costs. It might be many years more before it could pay off its debt and begin to earn a profit. Moreover, such a canal might never prove profitable and still be justifiable because of the economic development it precipitated. In a country where capital was scarce and other more attractive investment opportunities were abundant, private capital would not be forthcoming, and this kind of canal could only be accomplished by government enterprise or by private enterprise bolstered by very considerable government aid. The principal example of such a project, constantly before the legislators of New Jersey, was the great case of the Erie Canal. Its dominating purpose had been to open the Mohawk Valley and western New York to settlement and development, and it was not until later that its profitability became apparent. The importance of the Erie as an innovation which was widely imitated and thus contributed to the "canal mania" of the 1820s and 1830s has become a commonplace. The point here is the importance of the Erie example not merely as a stimulus to internal improvement in general but as a stimulus to improvement by state enterprise. The great benefit brought to New York State through the enlightened action of its government provided a powerful incentive to other state governments to "go and do likewise." The lessons of the Erie example were not lost upon New Jersey. One editor indicated his admiration and envy in quite glowing phrases:

The astonishing success which has attended the structure of Canals in the State of New York—the public benefits resulting from a safe and easy communication to the first commerical mart in the country— and the rich harvest which the State anticipates as the reward of its enterprise, has prostrated mountains into mole-hills, and given a universal tone to internal improvements. The impulse is sensibly felt in Pennsylvania and is operating like an invisible charm in the States of Ohio, Connecticut, and Massachusetts. New Jersey is catching the flame—her slumbering Samsons are beginning to awake—and the period is at hand when something efficient and in earnest will be done.[37]

On the other hand, it might be immediately and convincingly apparent that a particular canal would be a profit-making project virtually from the moment of its completion. Such was likely to be the case with waterways that served to facilitate existing trade or connected two major commercial centers. Here it was merely a matter of determining the volume of the trade likely to use the canal and balancing the resulting estimated revenue against the anticipated costs. If the results of this entrepreneurial calculus were particularly favorable, private capital might be expected to come forward and undertake the project. Or, as was repeatedly suggested in New Jersey in the case of the Delaware and Raritan Canal, the state might accomplish the task itself thus securing to itself the revenues to be obtained from the "resource" provided. In seeking specific illustrations for this type of canal, promoters resorted to English experience and drew attention to certain of the canals built in that country which had produced such fabulous returns. Thus, because of its relatively prompt and assured profit prospects, the exploitative canal might be the creature of either private or public enterprise whereas the developmental canal, because of its long-range and far more dubious prospects, could not be built without public action or public aid.

The New Jersey canals exemplified both of these kinds of canals. In common with the vast majority of canals built or projected in the United States during the period, the Morris Canal clearly fell into the developmental category while the Delaware and Raritan provides one of the few cases of an exploitative sort. Discounting the enthusiasm of its projectors in their estimates of traffic, prospects for the northern project were to depend upon the agricultural and industrial development of North Jersey and upon the substitution of anthracite coal for wood as an industrial fuel and also as a fuel for heating homes. It was expected that this substitution would be induced by the reduction in the price of coal that the canal would bring about by lowering the cost of transportation from the mines in Pennsylvania. There was no coal trade to be exploited at the time the canal was projected. The canal would create the circumstance out of which an exploitable trade would grow. Moreover, by revitalizing the iron industry and the agriculture of northern New Jersey the canal would similarly be able to exploit the ensuing trade. The point here is that none, or very little, of this traffic existed at the time. It would result from the economic development induced and made possible by the canal. Moreover, it was repeatedly asserted that even if the Morris never earned a penny of profit, the developmental effects would amply justify the state in undertaking it. Thus the Morris Canal fell into the category of developmental canals, canals which could only be built by governments or by private capital given massive government aid.

The Delaware and Raritan, on the other hand, was to capitalize upon the critical factor of its location between New York and Philadelphia and exact a tribute from the heavy trade passing between those two centers. As with the agitation for the Morris, projectors sought to unite the developmental argument with the exploitative in making their plea for popular support. However, in setting forth the arguments for the Delaware and Raritan, it is quite apparent that major dependence was to be placed upon the exploitation of the coastal and overland traffic. Purple passages picturing pleasant prospects for industry and commerce—"the canal lined with villages and dotted with factories"—were very largely an attempt to counteract the rightful apprehension of local merchants fearful of the destruction of their trade. The dominant argument was that of the traffic to be taxed by the canal. Thus the Delaware and Raritan may be placed in the category of exploitative canals open to either private or public enterprise.

It may be observed that New Jersey got what it expected to get from its canals. As anticipated, the developmental Morris accomplished a renaissance of the iron industry and aided materially in the industrialization of northern Jersey but turned out to be an unprofitable venture for its stockholders so far as canal operations were concerned. The Delaware and Raritan, on the other hand, though having little significance for the economic development of the state, was a most profitable undertaking both for its stockholders through dividends and for the state through dividends and transit duties. In some years income from the Joint Companies provided slightly more than half of annual state revenues.

As we have seen, the state of New Jersey provides a notable exception to the rule of state action in internal improvement. Neither of her major works received state aid; both were the exclusive product of private enterprise. The charters of the Morris and the Delaware and Raritan canals provided for no public subscription to stock, no grant of the state's credit, no financial aid whatsoever. In fact, in the case of the Delaware and Raritan Company of 1824 a charge of $100,000 was levied on the company, and in that of 1830 a gift of stock was required. The purpose of this study has been to examine the story of these two canals in an attempt to discover the peculiar forces and conditions which enabled the Legislature to provide for these tasks of internal improvement, huge for a state of New Jersey's meagre financial resources, without extending its credit to the projects.

All previous experience and contemporary advice militated against the resort to independent companies and urged state action. Prior companies chartered for internal improvement had met with failure owing largely to their inability to acquire the necessary capital. The success of the Erie as a public enterprise provided an outstanding example constantly brought to the attention of the legislators. The recommendations of experts consulted and commissioners retained were for public action. Moreover, the Legislature was reluctant to place such economically important and politically powerful projects in the hands of private enterprise; there was considerable fear of monied corporations, particularly in New Jersey where such companies seemed inevitably to end up in the control of out-of-state financial interests of New York or Philadelphia. Thus had its decision been taken on the basis of previous experience or contemporary counsel, the Legislature must have chosen to accomplish the projects itself and by this means ensure construction and retain control.

Nor did New Jersey refrain from direct action in internal improvement because of any ideological scruples of a laissez-faire nature; the moral right of the state to undertake such projects was never questioned. No fact emerges with more certainty from this study of the New Jersey canals than that throughout the long course of discussion and debate the right of the state to undertake them as public enterprises was not questioned. Opposition arguments made no mention of such an objection; the proponents of state action never found it necessary to defend against it. It was agreed that the state had the right and indeed the obligation to participate in projects for the promotion of economic activity. The question at issue was not whether the state could undertake or aid canals, but whether it should do so in the particular case at hand. Unencumbered by considerations of dogma or ideology the discussion could focus upon the purely pragmatic question of whether or not it would be expedient for the state directly to undertake the specific canal project. Clearly the mid-nineteenth century reaction against government intervention had not yet set in. Private capital had not yet become adequate to the task of internal improvement; hence government action was not impatiently and categorically rejected as an infringement of the rights of free enterprise but was eagerly accepted as essential to the economic development of a capital-short nation. Indeed, as a practical matter, the state seems to have been preferred to the monied corporation as an agent for the accomplishment of public tasks; big business was feared more than big government.

The principal determinants of New Jersey's policy toward its developmental canal were: the utility of the waterway to the region, the inability of private enterprise to secure the necessary capital, and the adamant opposition of certain groups to state action. The benefit of the improvement to the northern part of the state was not seriously questioned. Agriculture, industry, and commerce would be powerfully stimulated by the cheap transportation it would provide. However, it did not seem likely that, as a canal measure alone, it could interest sufficient investors, so that the mere creation of a private corporation having the right to construct the canal would not accomplish the purpose. On the other hand, the combination of southern agricultural counties and the proponents of the Delaware and Raritan prevented state action. Thus, as a canal only, the project would be doomed to failure since neither the state nor private enterprise would be willing to undertake it.

It is probable that the Morris Canal would never have been built had it not been for the tremendous enthusiasm for banking companies then sweeping New Jersey. The original canal party joined with the banking lobby to obtain the canal. The canal proponents seized upon the bank mania as a means of accomplishing their purpose without financial risk to the state; the banking interests seized upon the canal as a means of wresting a bank charter from a reluctant, even hostile, Legislature. By creating a private corporation the Legislature avoided any risk to the state and thus provided a means satisfactory to the opposition; by attaching banking privileges it ensured an ample subscription to stock. By holding out the promise of speculative banking profit New Jersey was able to attract private capital to invest in its developmental improvement and was by this means able to ensure its accomplishment without resort to public investment or mixed enterprise. It was only the bank mania of 1824–25 that enabled New Jersey to place her reliance upon private enterprise for completion of the Morris Canal. This enthusiasm for banking presented itself for only a very brief period, and by 1826 the speculative boom in banks was over. Thus the state was fortunate in making its contract at the peak of the market. Had it delayed by so much as one year, it is likely that the banking privilege, by then having lost much of its attractiveness, would not have provided sufficient inducement, and the company would have suffered the common fate of private companies created for canal purposes—failure due to inadequate stock subscription.

In contrast to the Morris, the Delaware and Raritan provides one of the few American illustrations of what we have called the exploitative type of canal. This was not a canal to connect a tidewater port with the interior; it would connect two major commercial centers. It would not open up a region for settlement and development; the region through which it was to pass and the regions which it would connect had already reached high levels of economic development. Nor would the coal trade provide a stimulus to industrial expansion within the area traversed by the canal; rather would it provide that stimulus in other areas —New York and New England primarily. For these reasons the Delaware-Raritan connection was rightly judged to be one of the most profitable canal opportunities afforded in the United States but was not expected to contribute greatly to the economic development of New Jersey. That these very optimistic profit expectations were not unfounded is indicated by the subsequent financial history of the Delaware and Raritan Canal Company. No other canal possessed the advantages enjoyed by it, and few canals could equal its financial success. This potential was clearly seen and constantly adduced by its supporters during the long period of its discussion. Thus the Delaware and Raritan as one of the exploitative canals could have been undertaken by either private or public enterprise.

This classification of the Delaware and Raritan provokes two major questions: (1) Why did the state not undertake the canal? and (2) Since the state did not, why was accomplishment by private means so long delayed? The answer to the first question should be prefaced by the remark that the state very nearly did undertake the project. In 1827 and again in 1828 bills for state action failed of passage by only the very narrowest of margins. This is enough at least to cast doubt upon the thesis that New Jersey refrained from an active role in internal improvement for powerful and overriding ideological reasons. Rather the refusal may be traced to the opposition to state action of powerful economic interest groups. The consistent hostility of the southern counties has been noted repeatedly as the story of the New Jersey canals has unfolded. Similarly the hostility of the proponents and later the proprietors of the Morris Canal has been indicated. Finally, in the last years of agitation for state action the opposition of the railroad party was apparent. Thus a combination of these various forces was able to prevent state action in the project despite the very liberal prospects such action promised.

The explanation of New Jersey's refusal to undertake the canal as a state measure does not account for the delay in its accomplishment by private means. As we have seen, two companies had been chartered, one in 1820 and another in 1824. The first company had failed for lack of capital. The depression of 1819 had adversely affected capitalists' expectations and had dried up the sources of capital. The company of 1824, coming in a period of speculative boom, had encountered no difficulties regarding subscription but had met with severe reversals in its attempt to obtain the permission of Pennsylvania for the diversion of Delaware River water. As negotiations dragged on, the boom of 1824 gave way to recession in 1826, and this caused the proprietors to lose heart and the company to be abandoned. It would seem that these difficulties could have been overcome had there been the prospect of obtaining a more liberal charter from New Jersey. Indeed, the New Jersey group in the management of the company urged such a course. However, a powerful element in the Legislature was reluctant to cede the profit opportunity which the Delaware and Raritan afforded to any private company. These legislators saw the canal opportunity as a resource which belonged to the people of the state and were unwilling to allow the tremendously profitable opportunity to be cheaply sold. Considerable fear was expressed for the political futures of legislators who might be party to such a "grab." Thus unable to undertake the project as state enterprise and unwilling to confer the opportunity upon private enterprise, the Legislature played the role of dog in the manger for nearly ten years. How long this situation would have continued can only be guessed. However, it seems unlikely that private capital would have allowed the unparalleled speculative boom of 1836 and 1837 to pass by without taking the bargain at pretty much the state's own terms. That this did not happen was, of course, the consequence of the coming of the railroad which threatened to destroy the value of the resource which the state possessed and forced the state to make the best bargain it could for its by then somewhat shopworn piece of goods.

To summarize, New Jersey was able to accomplish its major internal improvements without recourse to state enterprise only because of very unusual circumstances. It could ensure construction of its developmental project, the Morris Canal, only because of the short-lived speculative boom in banking of 1824. It was prevented from undertaking its exploitative canal, the Delaware and Raritan, by only the narrowest of margins and only by the opposition of powerful interest groups.

State action in internal improvement in New Jersey may be said to have foundered upon the twin rocks of sectional jealousy and conflicting economic interest. Similar local opposition had been encountered by the proponents of state action in other states. In New Jersey's neighbor to the east, the Erie Canal had been bitterly opposed by representatives of New York City, the Ontario counties, and the Delaware counties. This last acceded only when Clinton promised a state road as well as a branch canal to serve the region. To the west, the Pennsylvania mainline canal was similarly resisted by Delaware Valley interests and the Delaware Division Canal was quite frankly admitted to be merely a means of placating that section and gaining its support for the program. Such logrolling tactics were possible, of course, only when the resisting section had a project of its own which could be incorporated into the general plan. Had the southern New Jersey counties wanted roads or canals, such a *rapprochement* might have been accomplished, and a general program of internal improvement adopted. That they did not was probably the major factor in New Jersey's rejection of state enterprise and "steadfast adherence to the principles of 'laissez-faire.' "

The Economic Impact

IV. Cycles of Canal Construction

BY HARVEY H. SEGAL

In the United States the greatest undertakings and speculations are executed without difficulty, because the whole population are engaged in productive industry, and because the poorest as well as the most opulent members of the commonwealth are ready to combine their efforts for these purposes. The consequence is that a stranger is constantly amazed by the immense public works executed by a nation which contains, so to speak, no rich men. The Americans arrived but as yesterday on the territory which they inhabit, and they have already changed the whole order of nature for their own advantage. They have joined the Hudson to the Mississippi and made the Atlantic Ocean communicate with the Gulf of Mexico.

Alexis de Tocqueville [1]

Had De Tocqueville's remarks about American canals first appeared in 1840 instead of 1835, he would hardly have spoken of "the greatest undertakings" as being "executed without difficulty." For in the years following his visit canal construction in America was beset by enormous difficulties, bitter frustrations, and egregious failures. His premature admiration is, moreover, related to an enduring misconception concerning the course of American canal construction, that of the canal mania. According to it the American canal network, which by 1860 comprised more than 4,200 miles of line in twenty-three states, was the creation of a single canal movement. But in fact the ante-bellum canal network was the product of three distinct waves of construction activity which we shall call canal cycles. These cycles are unique phenomena in the development of the American economy whose origins can be traced to a particular conjuncture of technological, political and financial factors.

Measuring Canal Construction

There are two feasible approaches to the measurement of investment in canals. The first utilizes information on the number of miles of canals completed or placed into operation; the second is based upon annual construction expenditures or the value of construction work put into place. In this analysis we shall adopt the value approach for the following reasons.

Because of unresolved technological problems, difficulties in obtaining funds and other unforeseeable contingencies, canal lines were frequently completed or placed in operation long after the period of maximum construction activity—as measured by the absorption of economic resources—had been attained. These time lags varied widely among individual projects, and as a result the relationship between miles of canal lines completed and resource absorption, as measured by construction expenditures, is neither close nor consistent.[2]

Where the prisms of existing canal lines were enlarged in order to accommodate a larger volume of traffic, the mileage measure fails utterly to register the impact of construction activity. This deficiency of the mileage measure cannot here be ignored since the enlargement of the Erie and other lines of the New York State canal system accounts for more than twenty percent of the total investment in canals prior to 1861.

Although construction expenditures constitute a clearly superior measure of the volume of economic resources annually allocated to canal construction, H. Jerome Cranmer's investment series on which our analysis is based (see his paper on "Canal Investment, 1850–1860") will hardly satisfy the uncompromising purist. For a number of the smaller canals, which account for about twenty percent of the total investment in the period from 1815 through 1860, there are no extant records of construction expenditures. For those canals, estimates of annual expenditures were made by allocating their total costs over the elapsed construction period. These estimates, derived by techniques which are described in "Estimates of Canal Investment" (pp. 208–13), are subject to errors of unknown magnitude and direction. Fortunately, however, the years for which a high proportion of the canal investment is estimated are fairly well spread throughout the antebellum period so that the unknown errors of estimation should not strongly affect the time shape of the total investment series. This is shown in Figure 1 (p. 173), where the amount of investment that has been estimated in each year is indicated by the distance between the two time-series curves.

Contemporary accounts of annual construction expenditures are available for the remaining group of canals whose cost constitutes about eighty percent of the total investment. But the accuracy of the information derived from the contemporary records is subject to variation. For the fair-weather years, when the funds on hand were sufficient to remunerate the contractors for the value of construction work put into place, the contemporary records are generally reliable. Their usefulness, however, diminishes sharply in periods of financial difficulties when payments to contractors lagged far behind the pace of construction operations. Large debts on the public canals of New York, Pennsylvania, Ohio, and Indiana were incurred in this manner, especially during the period from 1839 through 1842. Where information on the volume of outstanding debts was available, we prepared special estimates of the value of construction work put into place by adding the estimates of outstanding indebtedness to the recorded expenditures. This method of adjusting the raw expenditure totals during the years of crises doubtless involves errors because it was necessary in several instances to allocate large balances of indebtedness over periods of several years. Such errors have probably distorted the time shape of the total investment series between 1839 and 1843 and again from 1856 through 1860,[3] but they do not affect our substantive analysis of canal cycles and their causes.

Canal Cycles: a General Analysis

Three long cycles or swings in canal investment are clearly discernible on Figure 1. The first and longest cycle runs from 1815

TABLE 1

Cycles of Canal Construction, 1815–60

	Duration (years)	Peak Year	Completed Mileage	Investment over Cycle (in millions of dollars)	Percentage of Total Investment
First Cycle, 1815–34	19.5	1828	2,188	58.6	31.1
Second Cycle, 1834–44	10.0	1840	1,172	72.2	38.4
Third Cycle, 1844–60	16.5	1855	894	57.4	30.5
Total for all cycles	46.0		4,254	188.2	100.0

Sources: Canal Investment: H. Jerome Cranmer, "Canal Investment, 1815–1860," Trends in the American Economy in the Nineteenth Century (Studies in Income and Wealth, XXIV; Princeton, N. J., 1960). Canal Mileage: Walter Isard, "A Neglected Cycle: the Tansport Building Cycle," Review of Economic Statistics, XXIV (1942).

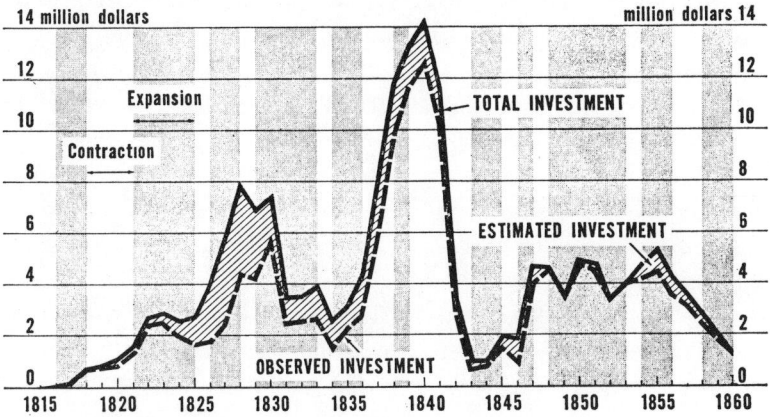

Figure 1. Canal Cycles, 1815–60

NOTE: Periods of expansion in general business activity are indicated by the shaded areas, and periods of contraction are indicated by the white areas.
SOURCES: See Chapter IV, Notes 2 and 4.

to 1834 and reaches a peak in 1828. A second cycle, with a very large amplitude of fluctuation, occurred between 1834 and 1844 with a peak in 1840; the last cycle ran from 1844 through 1860. Details concerning the relative importance of each cycle appear in Table 1.

We speak of *long* canal cycles because the duration of the swings in canal investment is much greater than that of the concurrent business cycles whose turning points are indicated on the chart. Periods of cyclical expansion are indicated by the shaded areas, and cyclical contractions are indicated by the white areas. Note that during the upswing of the first canal cycle alone there were four and a half cycles in general business activity and that the three long canal cycles encompass eleven business cycles.[4] In addition to their greater duration the canal cycles have turning points—peaks and troughs—that are largely independent of the turns in general business activity. Thus canal investment rose rapidly during the sharp cyclical contraction of 1836–38, and lagged the contraction of 1839–43 by a year.

Several questions come immediately to mind. Why were the canal cycles so long? Why were they largely independent of fluctuations in business activity? Why were there three canal cycles instead of a single long wave or perhaps a half dozen shorter ones?

In order to provide answers we shall construct an analytical model of canal cycles around three elements: (1) the technology of canal construction; (2) rivalry—a term which encompasses the commercial and political motives that impinged upon the decisions to build canals; and (3) finance or the availability of funds for construction.

Construction Technology. The reader who appreciates the complexities and difficulties of artificial waterways that required locks, culverts, retaining banks, and so forth, will readily perceive that their construction is a time-consuming process. Such is the case even where the most modern mechanical equipment is available. In America of the early nineteenth century, canal construction, though aided by the use of animals and blasting powder, was a time-devouring operation that was fraught with uncertainties. A contractor who agreed to excavate a section of canal line in a given period of time could rarely anticipate the problems that would arise once he delved into the subsoil structure. If fortune smiled, he would strike a soft, water-repellent clay, but frequently a less indulgent Providence plagued him with hardpan, porous sands, or solid rock. Time was also lost and costs were increased by inadequate supplies of hewn stone, timber and labor, especially on those canal lines which were cut through the lightly populated and economically underdeveloped frontier regions. Moreover, the severity of the winters in the northern states, where the bulk of the canal mileage was constructed, confined construction operations to the late spring, summer, and early autumn months.

The influence of these and other impediments is clearly reflected in long construction periods. In New York the 363-mile Erie Canal—the archetype of the public projects that were to follow—was under construction for nearly nine years, 1817 through 1825. Notwithstanding the experience gained in building the Erie, the 308-mile Ohio and Erie Canal, which ran from Cleveland to Portsmouth, required about the same construction time, from 1825 through 1833.[5] On the shorter projects the elapsed construction time far exceeded that which one would expect on the basis of length alone. Thus the 23-mile Chemung, one of the New York lateral lines, was under construction from 1830 to 1833; and the 60-mile Delaware Division line in Pennsylvania was under construction for nearly four years, 1827 through 1830.[6]

There is an important connection between long construction periods and canal cycles that emerges from a consideration of some simple arithmetic. Suppose that five canals are to be constructed, each of which will cost $10 million and require five years to complete. We shall assume in each instance that the financing is adequate, that there will be no delays in construction operations for lack of funds. Each project executed will then generate a five-year wave of construction expenditures. In the early months of operations outlays will be small as contractors assemble their work forces and acquire the requisite supplies of timber and stone. Thereafter outlays will rise rapidly to a maximum level and then slowly decline as the last details on the canal are completed. The resulting time-expenditure curve will be roughly bell-shaped.[7]

Let us now assume that each of our five canal projects have the same bell-shaped time-expenditure pattern and that the construction schedule calls for outlays of $1 million in the first year and $2 million, $4 million, $2 million, and $1 million in the second, third, fourth, and fifth years—or a total of $10 million for each canal. If all the projects are started in the same year a single five-year canal cycle which reaches a peak of $20 million in the third year and a terminal trough of $5 million in the fifth year will be generated.

Let us now suppose that instead of starting all the projects at the same time we stagger the starts by one year. Then a nine-year canal cycle reaching a maximum of $10 million in the fifth year will be generated. As we lengthen the stagger period the cycles become longer and the amplitudes of their fluctuation diminish. When the stagger period is equal to or greater than the construction period, the long cycles disappear and are replaced by a series of short cycles. The relationships between the stagger

period, the number of canal projects, and the duration of canal cycles may be expressed algebraically in the following manner:
Let

  pt = the required construction time for each canal
  n = the number of canals
  s = the length of the stagger period
  ct = the length of the canal cycle

Then

  $ct = (pt - s) + ns$

The long cycles resulted from the clustering of canal starts which occurred when n, the number of projects, was large and s, the stagger period, was very short. But why did these clusters occur? The answer lies in the rivalry among the principal seaboard cities for the trade of the Ohio River valley.

The Role of Rivalry. The construction of the Erie Canal—which was heralded as a success even before its completion—did not initiate "the canal movement," but it was the demonstration which dramatically transformed existing plans for internal improvement into operational realities. The force which conducted and amplified the electrifying impulse of the Erie success was the rivalry for the trade of the Ohio River valley. New York scored the first great success, and in 1826 the Pennsylvanians rose to the challenge and embarked upon the construction of a mainline between Philadelphia and Pittsburgh. Two years later, in 1828, the ill-fated Chesapeake and Ohio was launched. West of the Appalachians work began in 1825 on the Ohio and Erie Canal which completed the last link of an all-water route between New York and the Ohio River. The construction of those four interregional trunk lines involved an initial investment of some $30 million, a total which was augmented considerably by the addition of lateral or branch lines.

Not all of the principal interregional lines were under way by 1828 or even during the first cycle of canal construction. The demonstration effect of the Erie was potent, but its operation was subject to lags. In Virginia the impact of the Erie success was delayed because the state government was committed to the execution of an essentially intraregional project. Beginning in 1795 a private company, later taken over by the state, attempted to construct a canal along the James River. Although a water and turnpike route to Ohio River via the Kanawha was planned, construction operations until 1835 were limited to less than forty miles of canal line in the vicinity of Richmond.[8] In 1835 the James River and Kanawha Company was chartered, and, with generous assistance from the state government, it embarked upon a fourth trans-Appalachian route. On the western side of the mountain barrier, it was the immaturity of the newly formed state governments rather than commitments to established enterprises that caused the delay in launching interregional canal lines. Indiana, aided by a federal land grant, authorized the first sections of the great Wabash and Erie Canal in 1832, and the Illinois and Michigan Canal was placed under construction in 1836.

Taken alone, the seven major interregional canals—the Erie, the Pennsylvania Mainline, the Chesapeake and Ohio, the James River and Kanawha, the Ohio and Erie, the Wabash and Erie, and the Illinois and Michigan—would in the course of their execution have produced long canal cycles, but their amplitudes would have been much smaller than those of the cycles shown in Figure 1. For, in addition to the great interregional canals, there was an important group of local or intraregional canals whose construction intensified the degree of clustering and so increased the amplitude of the cycles of total canal investment.

The motive which shaped the decisions to build intraregional canals was a desire to enlarge the scope of local trade. Both the Middlesex Canal in Massachusetts and the Santee in South Carolina were constructed in order to attain such limited objectives long before the Erie was launched.[9] But it was the success of the Erie that precipitated a great upsurge of interest in local projects.

The intraregional canals fall into two categories. First, there were the relatively short lines, built largely by private interests, that connected commercial centers or facilitated the transportation of important raw materials. In this category we place the coal canals of New York and Pennsylvania—the Delaware and Hudson, the Lehigh Coal and Navigation, and the Union Canal

—as well as local trade lines, such as the Farmington in Connecticut, the Blackstone in Rhode Island and Massachusetts.

A second and quantitatively more important group of canals, whose planning and execution were profoundly affected by political forces, were those built by the states as branches or laterals to the public trunk lines. The decisions to build branches were in almost every case the result of powerful pressures exerted in the legislatures by the "unimproved" counties, those which were by-passed by the trunk-line canals. Even before the Erie was completed, there were loud demands for the construction of lateral lines, and in 1825 the New York Legislature authorized surveys for seventeen projects whose execution would have added more than 1,200 miles to the existing network. A determined band of financial conservatives in the New York Legislature limited to three the number of laterals projects undertaken in the years before 1835, but its resistance crumbled when it became necessary to enlarge the Erie Canal, whose original prism, only four feet deep and thirty feet wide at the surface, proved inadequate in the face of the rapidly growing barge traffic. The political price of enlarging the Erie—and later the Champlain and Oswego canals—was an acquiescence to the demands for additional laterals. In Pennsylvania, as seen in Chapter II, the "branch men," the proponents of lateral lines, wielded such great influence in the planning stage that in 1826 the state embarked upon the simultaneous construction of a difficult 395-mile Mainline between Philadelphia and Pittsburgh and 314 miles of largely unconnected lateral lines. During the first canal cycle, public construction in Ohio was confined to two trunk-line routes—the Ohio and Erie Canal, connecting Cleveland with Portsmouth on the Ohio River, and the Miami Canal, which linked Cincinnati with Dayton. But in 1836, when the state embarked upon the extension of the Miami Canal to Toledo on Lake Erie, it was necessary to authorize also the construction of nearly two hundred miles of laterals.

Although the rivalry between the proponents of trunk and branch lines centered mainly about the public canals of New York, Pennsylvania, and Ohio, its effects upon the total volume of canal investment should not be underrated. Canal investment in those three states accounts for fifty-four percent of the national total in the period from 1815 through 1860, and for nearly fifty-eight percent of national investment over the second and third canal cycles, 1834 through 1860.[10] In fact, the unsatisfied demand for lateral canals in those three states, together with the need for extending or enlarging the existing trunk lines, accounts in large part for the emergence of the second canal cycle.

Finance. The role of finance in generating long canal cycles is both obvious and important. If the spirit of emulation had not been nourished by flows of funds, there would have been no clustering of canal starts and hence no long canal cycles.

Of the $188 million invested in canal construction before 1861 some $137 million or seventy-three percent of the total was provided by state and municipal governments. The bulk of the government funds, about $115 million, was channeled into the public canals of New York, Pennsylvania, Ohio, Indiana, Illinois, and Virginia. An additional $22 million were made available by governments in the form of stock subscriptions and loans to mixed enterprises, of which the Chesapeake and Ohio and James River and Kanawha canals were the most important. (See "Public Funds in Canal Construction, 1815–60" [pp. 213–15] for detailed estimates of those subscriptions and loans.)

More than ninety percent of the funds provided by governments—or not much less than $127 million [11]—were raised through loans. Our knowledge of the canals that were built exclusively by private capital is limited, but the available evidence suggests a heavy reliance upon loans. It is therefore probable that at least eighty percent of canal construction funds—or some $150 million—was raised through loans.

At least three fourths of the $150 million of canal loans were obtained through the sale of bonds to financial institutions—bankers, insurance companies and brokerage houses. Some $62 million of state government bonds, which provided nearly a third of the total construction funds, were purchased by foreign banking houses, principally in Great Britain.[12]

These facts bear significantly upon the question of the availability of canal funds. Private risk capital in the United States, at least before 1845, was scarce,[13] and none of the canal building states was willing to raise construction funds by levying new taxes. As a result, a heavy reliance was placed upon borrowing through the sale of bonds in the domestic and foreign money markets. The ability to raise funds in this manner was proximately determined by the reserve position of the banks, that is, by the ratio of specie held in banks to note and deposit liabilities. During periods of high optimism and satisfaction with current reserve positions, banking institutions—commercial banks, brokerage houses, insurance companies—in the United States and Great Britain readily purchased canal bonds, which could then be shifted to individual investors. But those institutions were extremely sensitive to internal and external disturbances. A fear-borne desire by the public to increase their holdings of specie, a shift in the balance of payments occasioned by a crop failure, or a change in the relationship between internal and external price levels could within a very short time cause a seemingly inexhaustible supply of long-term funds to vanish.[14]

Although foreign loans accounted for less than a third of the total canal funds raised before 1861, their significance was far greater than this over-all figure would suggest. Most of the foreign loans were extended between 1825 and 1840 and were heavily concentrated during the vigorous expansion phase of the long second canal cycle—from 1835 through 1840—when, directly or indirectly, they provided more than ninety percent of the funds invested.[15] Foreign loans were also important, even in periods when they supplied much smaller proportions of the construction funds, because the expectation of their availability, even when illusory, provided an incentive for domestic banking institutions to purchase canal bonds for resale abroad. On the other hand, a tightening of the foreign money markets, for whatever cause, was interpreted by the domestic banks as a danger signal. In anticipation of impending difficulties they would curtail their loans and, in particular, their purchases of long-term canal bonds, and thus raise their reserve ratios.

We are not suggesting that the state-chartered banks of the

ante-bellum United States—or the institutions that dominated the London money market—faithfully adhered to latter-day textbook maxims for operation under a gold specie standard. All that we are claiming is that the markets for long-term funds under the gold standard of the period were delicately balanced and perilously vulnerable to shocks. As a result the supply of loanable funds, upon which canal construction was so heavily dependent, was subject to sudden and violent fluctuations.

The Model. As a first approximation, we shall endow American economic and political life in the early nineteenth century with a rationality that has hardly been attained in the mid-twentieth. Let us assume, first, that the conflicting sectional interests have been somehow composed and that decisions to build canals are the sole province of apolitical, federal technicians. Second, let us assume that the federal government can obtain canal funds without delay either through taxation, loans, or a resort to the printing presses. In such circumstances, a conservative board of federal engineers might embark upon a long-term program of staggered canal starts. In that case, either a series of short canal cycles or a relatively constant level of construction activity over a long time period would result. A less cautious group of engineers might venture upon a program of simultaneous construction over all feasible canal routes, and there might then ensue a single wave of construction activity. In any case, it is extremely unlikely that the three long cycles which we have observed would have occurred within the context of the assumed conditions.

We shall now disband the bureau of federal engineers, thus consigning construction decisions to the political jungle, while at the same time retaining the federal government's ability to finance all undertakings without delay. Now the decisions to build canals are determined by the relative political power of states and regions, the prestige and personalities of the contending political figures, changes in public opinion, and a host of other factors which we need not enumerate. By making any one of a number of equally plausible suppositions concerning political or technological developments, the appearance of a series of long canal cycles will logically follow. All that is really essential is that there be substantial time lags between the clusterings of canal starts. Each of these cycles will run its natural course, uninterrupted by financial difficulties of any sort.

By stripping the federal government of its power to raise canal funds and transferring that function to the state governments, we eliminate the last unrealistic assumption and confront the stark realities of economic life in the early nineteenth century. Financing canal construction now involves the sale of bonds by competing borrowers who converge simultaneously upon the imperfectly functioning money markets, which are characterized by a high degree of instability on the supply side. Any disturbance, whether caused by international political frictions or the recurrent business cycles, results in a rapid diminution in the supply of loanable funds and difficulties in sustaining canal construction.

When our model is applied to the period under investigation, the following generalizations emerge.

The first canal cycle (1815–34) is largely a reflection of the great Erie success and the undertakings in Pennsylvania, Ohio, and Maryland which followed rapidly upon its heels. The techniques of floating canal bond issues both at home and in the London market were rapidly perfected, and, largely as a result of this development, construction activity during the first canal cycle was virtually uninterrupted for want of funds. Construction activity turned downward after 1829 only because a preponderance of the projects undertaken earlier were nearing completion.

A number of projects, both interregional and intraregional, were surveyed but not executed during the first cycle because of the immaturity of western state governments and the success of conservatives in repelling the demands for additional laterals in the eastern canal states.

After a brief pause, a second and far more vigorous upturn in canal construction commenced. The second cycle (1834–44) it should be noted, was closely linked to the international inflation of commodity prices, especially that of cotton, and to the frenetic burst of developmental activity in the United States, accompanied by wide-spread financial speculation. The ease with which loans were obtained in the London and other foreign money markets permitted the commencement of numerous lateral canals in the established canal states as well as the execution of major interregional lines in the west. As a result, the volume of construction was expanded vigorously between 1834 and 1840, the upward trend being completely unaffected by the cyclical contraction of 1836–38 for reasons which will be presently discussed. The end of this great expansion came in 1840 when, as a result of the cyclical downturn which had occurred in the previous year, the stream of foreign capital ceased to flow. This development precipitated a crisis in the affairs of canal states. Notwithstanding a number of financial improvisations, canal construction had all but ceased in most states by 1842. The financial crisis resulted in an epidemic of defaults on canal loans which had a devastating effect upon the credit standing of the states. This second, or interrupted, canal cycle came to a close with nearly $60 million of work unfinished.

After a period in which the fiscally improvident canal states regained solvency by a painful recourse to taxation, work was once again resumed on the abandoned projects, giving rise to the third and last canal cycle of the ante-bellum period. This cycle was an echo phenomenon, involving unfinished construction in a period when there were few grounds for believing that artificial waterways could effectively compete with railroads. Unlike its predecessors, the third canal cycle was principally financed from domestic sources.

The First Canal Cycle, 1815–34

During the first, or uninterrupted, canal cycle, a large portion of the United States canal network was rapidly completed with a minimum of difficulties. The figures in Table 1 (p. 172) indicate that the first cycle accounts for more than half of the mileage and thirty-one percent of the total investment outlays in the ante-bellum period. In appraising its significance, one should bear in mind that the volume of investment is estimated at current prices. Constant dollars estimates are clearly superior for pur-

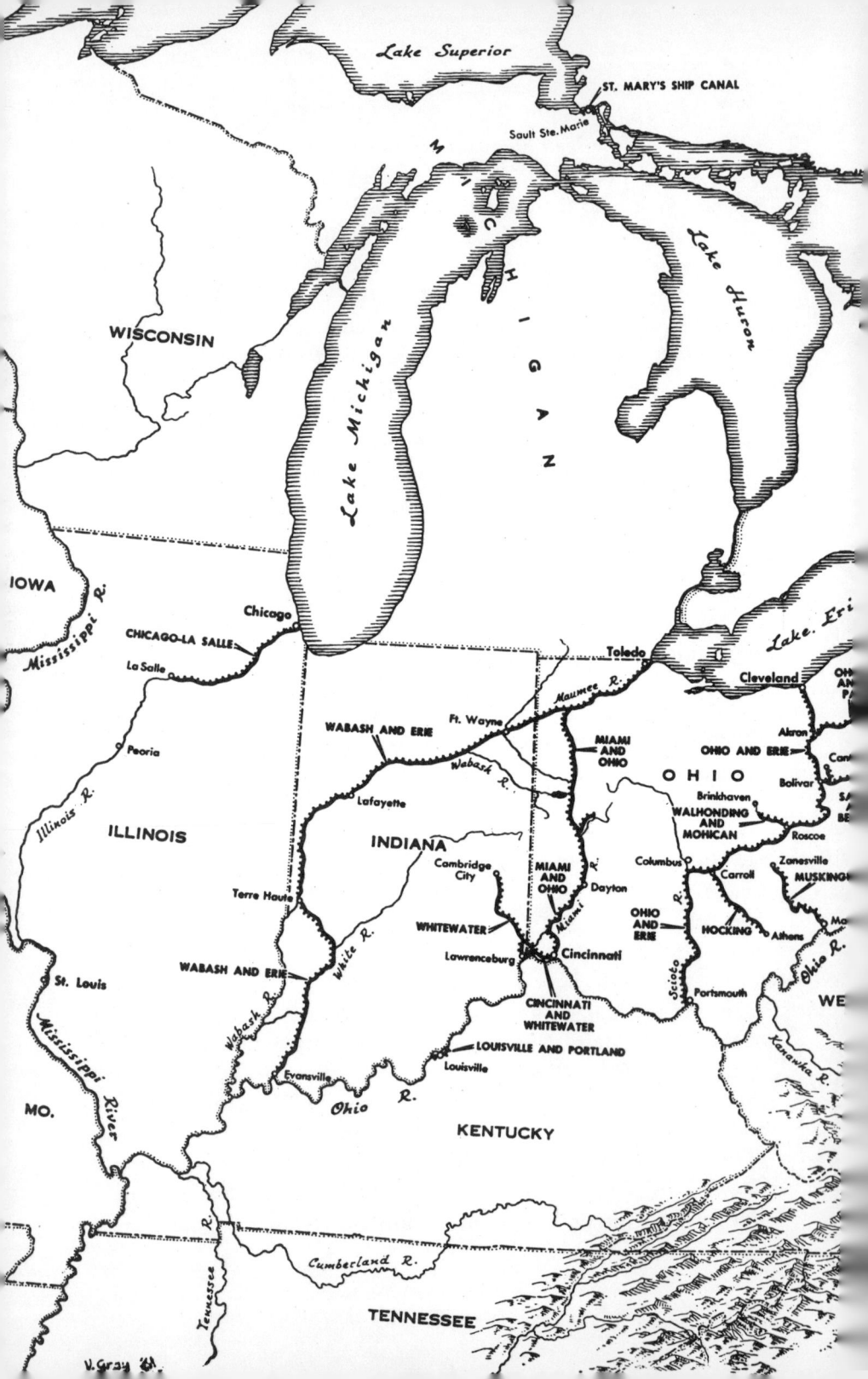

poses of comparison over long time periods, but the information necessary for the preparation of construction cost indices for canals is very difficult to obtain.[16] We believe, however, that average construction costs rose sharply between the first and second canal cycles—perhaps by as much as thirty-three percent. It is possible, therefore, that the first canal cycle accounted for about forty-seven percent of the real volume of total investment.

Among the impressive achievements of the first wave of construction activity were the Erie and Champlain canals of New York, the Mainline of the Pennsylvania system, and the eastern trunk line of the Ohio state network, the Ohio and Erie Canal. To these one should add a number of private or mixed canal enterprises, the most important of which follow. In New England, there were the Blackstone, the Farmington, and the Hampshire and Hampden. Among the important lines in the Middle Atlantic states were the Delaware and Hudson, the Morris, the Delaware and Raritan, the Schuylkill Navigation, the Union, and the Chesapeake and Delaware.

Publicly financed projects—state canals and mixed enterprises that depended heavily upon governmental assistance—clearly dominated the first cycle, accounting for $41.2 million or 70.3 percent of the total investment. The shape of the first canal cycle, shown in Figure 1, clearly reflects the timing of these public projects. Work on the New York canals was commenced in 1817 and that on the James River Canal of Virginia in 1819. By 1823 construction activity in New York and Virginia had passed its peak and aggregative investment would have fallen sharply downward had it not been bolstered by the commencement of the Ohio canals in 1825 and those of Pennsylvania in 1827.

Financing Construction in the First Cycle. It is safe to say at least eighty percent of the nearly $59 million that were invested in canals during the first cycle was obtained through loans. Our certitude, however, diminishes when we inquire about the lenders. Were they individual investors or financial institutions? What was the role of foreign capital?

In a recent study of the Erie Canal, Nathan Miller [17] has shown that the first canal loans were made principally by small investors, and that large institutional investors were not attracted until the success of the canal seemed assured by the substantial tolls collected on the completed sections. This shift from individual to institutional investors appears not to have been repeated in the case of other public ventures. From the beginning, the Ohio Canal Fund Commissioners borrowed in the New York money market, and of the first $4.5 million of Ohio canal bonds issued, $870,000 were sold to the firm of Prime, Ward and King, the American correspondent of the Baring Brothers, while even larger blocs went to banks and other financial institutions. In Pennsylvania, a single lender, the Bank of Pennsylvania, purchased more than two thirds of the state canal bonds issued up to the end of 1829. The officials of the Chesapeake and Ohio Canal Company, who had to sell state and municipal bonds that were transferred to the company in payment for stock subscriptions, also relied heavily upon institutional investors. One reason for participation of banks was the rapid retirement of the federal debt, which declined from an average of $90 million in the early 1820s to a low of under $5 million by 1834.[18]

The question of foreign-investor participation during the first canal cycle poses great difficulties and compels an even greater reliance upon conjectures and inferences. One point, however, about which we may be reasonably certain is that in the early stages of the first cycle the state canal officials were looking with confidence to the London money market for loans. In 1823 Micajah T. Williams, an Ohio Canal commissioner, wrote to John T. Champlin, president of the Farmers' Fire Insurance and Loan Company of New York City, seeking his opinion as to the prospects for floating a $2.5-million canal loan. Champlin's optimistic reply contained the following remarks on foreign loans.

The contractors of stocks look to the London market also, for sales to great extent oftentimes. Indeed of late, the stocks in this country are a pendulum to regulate foreign bills of exchange, and are negotiated as a substitute to the latter.[19]

The question, therefore, is not whether foreign capital played an early role, but rather what was the magnitude of that role. Once again we are confronted by a paucity of evidence. There are strong grounds for assuming that the Bank of Pennsylvania and other institutional purchasers of canal bonds periodically disposed of their holdings abroad, perhaps in settlement of debts due their foreign correspondents. Leads concerning the over-all importance of these and other foreign transactions come from Nathan Miller, who estimates that more than half of the outstanding New York canal bonds were in the hands of foreign investors by 1829, and from a contemporary source, which places one half of the Pennsylvania bonds abroad by 1832.[20] If we allow for a sharp increase in the foreign holdings of New York bonds after 1829, we reach the conclusion that about $13 million of New York and Pennsylvania bonds were in foreign hands by 1832.[21] If we add to this $2 million as an estimate of the amount of other state canal bonds held abroad, and we arrive at a total of $15 million, roughly a third of the public funds and more than a quarter of the total funds invested.

An Unbroken Cycle. In the first canal cycle, few of the problems encountered can be traced directly to difficulties in obtaining funds. A number of factors contributed to the relative ease with which the construction of nearly 2,000 miles of canal line was financed. Until the early 1830s when the bulk of the construction work had been completed, the monetary system of the United States was unscathed by partisan political conflict, and the general confidence which prevailed throughout the 1820s was conducive to the smooth functioning of the money markets. Moreover, the years from 1821 until 1833 were characterized by a remarkable degree of cyclical stability. We know relatively little about the recessions of 1825 and 1829, but neither appear to have been severe and only the latter created difficulties, albeit not serious ones, in the markets for long-term canal funds. The more severe contraction of 1833–34, which was exacerbated by the uncertainties growing out of the dispute between Andrew Jackson and Nicholas Biddle over the Second Bank of the United States, occurred too late in the first cycle to cause much harm.

Notwithstanding this conjuncture of favorable economic factors, there were two notable instances where difficulties were encountered in obtaining funds during the first cycle. The first involves the Chesapeake and Ohio, a canal whose long construction period—from 1828 until 1850!—was characterized by an unbroken series of delays, misfortunes, and frustrations. Its funds exhausted, the company in 1831 was compelled to reduce the volume of construction activity at a stage in the canal's progress when it should normally have been vigorously expanded. During the recession of 1834, the Chesapeake and Ohio management, harrassed once more, initiated a practice to which they had frequent recourse in later years. They issued some $129,000 of canal "scrip" in small denominations. This irredeemable currency circulated until the company was able to retire it.

In the recession of 1829 the Pennsylvania canal fund commissioners, who had previously floated nearly $5 million of long-term loans with ease, found it impossible to dispose of a new $2.9-million five-percent issue and were compelled to resort to temporary financing at higher rates. They attributed their failure to obtain permanent loans, not to the tightness of the money markets, but to the opposition of the private directors of the Bank of Pennsylvania, who constituted a majority of the board, notwithstanding the state's sixty percent interest in the capital stock of the bank. Revenge for an angry Harrisburg Legislature came in the following year when the bank's special charter expired. According to the terms of the new charter, the bank was compelled to extend permanent improvement loans up to $4 million and temporary loans of as much as $1 million annually. In the summer of 1834, when funds were hard to come by, the state exercised its prerogative by borrowing nearly $2.3 million. However, a concession, which foreshadowed later practices, was granted; the bank was permitted to make payments in its own notes. By scattering notes of small denomination throughout the sparsely settled areas where the bulk of the construction work was concentrated, the bank protected its reserve position, postponing the day of reckoning when other banks or individuals would return its notes for redemption.

The Second Canal Cycle: Expansion and Crisis

In the second cycle, canal investment soared from relatively low levels in 1834 to a peak of more than $14 million in 1840,[22] only to drop to a level of less than $100,000 in 1844. The violent upsurge and precipitous decline of canal investment were not isolated phenomena. In order to comprehend them, we must first consider the general economic developments of which they were a part.

Unlike the decade which preceded it, the period 1834 to 1844 was characterized by an extremely rapid pace of economic development in the western United States and a high degree of cyclical instability. Among the salient characteristics of the boom, which persisted with varying degrees of intensity until 1839, were the inflation of land and commodity prices, a sharp expansion of the money supply, together with the uncertainties growing out of the banking conflicts, and the great influx of British capital.

Following the cyclical contraction of 1834, there was a land boom and a sharp rise in the prices of agricultural commodities, especially that of cotton. Land speculation, which reached its dizziest heights in 1835 and 1836, centered around the extension of cotton cultivation to the states of the old southwest and the rapid settlement of the lands in the vicinity of the Great Lakes. Many aspects of this movement, which must have involved an enormous increase in land wealth, are as yet imperfectly understood. The westward expansion of cotton cultivation was apparently associated with actual or anticipated increases in the world demand for cotton and the concomitant rise in its price. The land boom in the old northwest was accompanied by a rapid influx of population which was partially induced by transport improvements, of which the Erie Canal was the most notable.[23] In the course of these developments, the proceeds from public land sales rose from an annual average of $1.1 million in the decade of the 1820s to $7.8 million in the 1830s, reaching a peak of nearly $25 million in 1836. Agricultural prices, which had been stable throughout the 1820s, rose sharply after the 1834 recession and, except for a sharp break in 1837, were maintained at a very high level until the contraction of 1839–43.[24]

The speculation in land and commodities was accompanied by a vast monetary expansion. The money supply—bank notes plus deposits—increased by fifty-five percent in the five-year period 1834–39.[25] This expansion commenced when the federal charter of the Second Bank of the United States expired, creating a vacuum that was rapidly filled by state-chartered banks, the number of which increased from 501 in 1834 to 901 in 1840.[26] Among the significant factors in this expansion was the granting of a Pennsylvania charter to the Second Bank of the United States early in 1836. Under the brilliant but erratic leadership of Nicholas Biddle, the Bank subsequently played an important role in the negotiation of foreign canal loans. The passage of a Free Banking Act by the New York Legislature in 1838 also facilitated canal construction since, under its terms, the banks could issue notes when equivalent amounts of state bonds were deposited with the comptroller.

The last and most important economic development was the enormous influx of foreign capital. According to recent estimates by Douglass C. North, the total foreign indebtedness of the United States, which fluctuated about an average level of $85 million in the 1820s, rose rapidly to more than $297 million in 1839.[27] This inflow of capital, which did not abate until the fateful cyclical downturn of 1839, balanced the large import surplus of the United States during most years in the decade of the 1830s and prevented a deterioration of the foreign exchanges. So long as it continued the domestic banking institutions could absorb canal bonds with comparative safety.

Since British investors provided the bulk of the long-term loans that were extended during this period, it is appropriate that we appraise their role in the American boom. R. C. O. Matthews, who has recently written with great insight into this period, inclines toward the view that Britain played a passive role in the American boom. "The bonds were bought up as they came forward," he concludes, "but the initial propelling force lay in internal development in the United States itself." [28] On this judgment we must demur. The phrase, "the bonds were bought up as they came forward," ignores the role of the Anglo-American bankers whose well-advertised eagerness to purchase state bonds encouraged the undertaking of projects of internal improvement and so were in part responsible for the bond issues.[29] The truly passive party was the individual British investor who found the American state bonds quite attractive during a period in which the yield on consols (British government perpetuities) was declining.[30]

We estimate that about $43 million, or sixty percent, of the total investment during the second cycle were raised through foreign loans.

Accelerating the Pace of Construction. A vast expansion of construction activity carried the volume of canal investment from a low of $2.1 million in 1834 to the ante-bellum peak of $14.2 million in 1840. The impetus came largely from a series of legislative decisions affecting the public projects,[31] which in the second cycle absorbed $57.3 millions or nearly eighty percent of the investment. These decisions, which tended to cluster around 1836, frequently involved significant departures from previous attitudes or practices.

NEW YORK. Although the New York decision came later than the others, it involved the most important single project in the ante-bellum period, the enlargement of the Erie Canal. The Democrats under Governor W. L. Marcy were committed to a conservative canal program. The Erie enlargement was to be financed solely from surplus canal revenues, which, according to the estimates, would provide $1 million annually over a twelve-year construction period. Work on the enlargement began in 1834.

With the Whig victory under William H. Seward in the autumn of 1837, there was a radical change in policy. Early in 1838, the Whig majority of the Ways and Means Committee declared that the surplus canal revenues were sufficient to sustain and retire an internal improvements debt of $40 million without recourse to direct taxation. They urged an immediate acceleration of construction activities, not only on the Erie enlargement, but also on the Black River and Genesee Valley lines, work on which had been commenced in 1836. Within a month, in April of 1838, the Legislature authorized a $4-million loan for the Erie enlargement. The way was now cleared for a rapid expansion of construction outlays in New York that reached a peak of $5 million in 1840.

PENNSYLVANIA. After the completion of the 395-mile Mainline, together with 314 miles of laterals, the Pennsylvania improvements debt in 1834 amounted to more than $20 million.

Notwithstanding this large debt, Pennsylvania embarked upon a new construction program when it granted a state charter to the Second Bank of the United States in February, 1836, by "An Act to repeal the state tax on real property, and to continue and extend the improvement of the state by railroads and canals, and to charter a state bank, to be called the United States Bank." Under its terms the state was to receive a charter "premium"— an outright gift—of $2 million for purposes of internal improvement, and, in addition, the bank was committed to extend to the state long-term improvement loans of up to $6 million, at four percent interest, and temporary loans of as much as $1 million for twelve-month periods. In the next two years Pennsylvania embarked upon the construction of some 225 miles of new canals, the estimated cost of which exceeded $7 million.[32]

OHIO. Between 1833 and 1836 canal construction in Ohio was confined to a thirty-three mile extension of the Miami Canal north of Dayton. Construction funds were derived solely from the sales of federal lands that had been granted to the states in 1828, and the scale of operation was limited. A marked expansion began in March of 1836 when the Legislature voted to authorize the construction of 194 miles of lateral lines. These, together with the Miami extension and the Ohio section of the Wabash and Erie Canal, added 408 miles to the Ohio network at a cost of $9 million.[33]

INDIANA. In 1827 Congress granted Indiana lands to aid in the construction of a canal to connect Lake Erie and the Ohio River.[34] When completed in 1852, the 467-mile Wabash and Erie Canal, which connected Toledo with Evansville, was the longest artificial waterway in the United States. Construction operations on the northeastern sections of the line were under way by 1832, but very little was accomplished before 1836.[35] In January of that year the Indiana Legislature passed what became known as the "Mammoth Internal Improvements Bill," since it authorized loans of $13 million for internal improvements. Of this total, $5 million were appropriated for the completion of the Indiana sections of the Wabash and Erie Canal [36] and $1.4 for the construction of the 76-mile Whitewater Canal, connecting Cambridge City and Lawrenceburg in eastern Indiana.

ILLINOIS. Congress authorized a grant of land to assist in the construction of a canal between Lake Michigan and the Illinois River in 1822, but nearly thirteen years passed before the Illinois Legislature committed the state to undertake the $6-million project in January of 1836.[37] Among other causes for the long delay was the stubborn resistence to any proposal that involved the incurrence of a state debt.

MARYLAND. Before 1835 Maryland's role in financing the Chesapeake and Ohio Canal was limited to a modest $500,000 stock subscription. As a result of the pressures generated in an internal improvements convention, the Legislature in 1835 authorized a $2-million loan for the hard-pressed company. A second, far more effective improvements convention, resulted in the "The Eight Million Act." Passed in a special session in May, 1836, it authorized a $3-million subscription to the stock of the Chesapeake and Ohio Company. This was followed by a $1.65-million loan in 1839. Altogether the state of Maryland authorized $6.65 million in aid to the Chesapeake and Ohio during the second cycle, and a $1-million loan to the Susquehanna and Tidewater Canal Company in 1839.

VIRGINIA. Between 1819 and 1835 the state of Virginia spent about $3.3 million on the James River Canal, principally in the vicinity of Richmond.[38] The groundwork for a far greater effort was established in January of 1835 when the State Legislature substantially increased its stock subscription to the James River and Kanawha Company. In the twenty years that followed, the company, with very liberal assistance from the state, built 196.5 miles of canal line between Richmond and Buchanan at a total cost of nearly $8.9 million.

Financing the Expansion. In negotiating the sales of the state bonds which provided about three fourths of the total of funds raised during the second canal cycle, the large financial institutions—banks, insurance companies, and brokerage houses—played a dominant role. The Bank of the Manhattan Company was active in the market for New York bonds. A substantial share of Ohio bonds was purchased for resale in London by the Ohio Life Insurance and Trust Company and the New York firm of Prime, Ward and King, both of which drew against an open line of credit with Baring Brothers.[39] The Chesapeake and Ohio Company relied first upon George Peabody and then upon Baring Brothers in disposing of large blocks of Maryland bonds.

Among the institutions which operated in the state bond market, the Bank of the United States was preeminent. Between 1836 and its collapse in 1841, the Bank and its affiliate, the Morris Canal and Banking Company, purchased nearly $30 million of state improvement bonds of which about $17 million were issued for canal construction.[40] The alacrity with which Biddle and his associates acquired this formidable block of securities can in large part be explained by the cotton operations in which they were engaged during the period 1837 through 1840.[41] Working through affiliated brokerage houses, Biddle and his associates on three occasions cornered substantial shares of the cotton crop and held out for a price rise. The success of these ventures rested heavily upon the ease with which credit could be obtained from London, and, for that purpose, the American improvement bonds, to which the faith and credit of the states was pledged, were well suited. After a brilliant initial success these operations ended badly when, in the autumn of 1839, it was no longer possible to sell the state bonds or pledge them for loans.[42]

As the debts mounted, the fiscal position of the canal states became precarious. Only the New York canals produced substantial net revenues, but even those became inadequate to carry the debt charges. A general abhorence of "direct taxation," that is, taxes on real property, caused all the canal states to engage in the dangerous practice of servicing their debt, with the proceeds of loans. Taxation, as we shall see, was resorted to with extreme reluctance only when circumstances permitted no other course.

Difficulties in financing canal construction were first encountered late in 1836 and during the celebrated financial panic of

1837. The trouble began when the London market, which had been buoyant since 1833, began to tighten in the spring of 1836. For some months prior to that time the directors of the Bank of England had been restive over the volume of bills growing out of the Anglo-American trade. A large number of discounted American bills, based upon state securities, were being offered as collateral for brokers' loans, and the large-scale transfers of funds was accompanied by a loss of specie. In July, when Jackson's "Specie Circular" created some alarm, the directors raised the bank rate to four and a half percent for the first time since 1827. Yet the specie drain continued, and in September the rate was again raised. At that time the directors made it known that they took a dim view of American securities as collateral for loans or remittances against drafts.[43]

Notwithstanding the severe tightness in the London and American money markets, the volume of construction activity along the new canal lines, as indicated on Figure 1, rose sharply, from $4 million in 1836 to $8 and $11 million in 1837 and 1838. This contracyclical expansion of canal investment was made possible by the large quantity of funds that was on hand when the crisis arrived and by the rapid recovery in the money markets that followed early in 1838.

Most of the problems encountered during the crisis of 1837 were surmounted without great difficulty. The New York Canal Fund Commissioners raised the sterling exchange required to meet the interest on the foreign-held bonds by "lending" a block of new issues to a friendly New York City bank to be used as collateral for loans abroad. Pennsylvania remitted depreciated dollars to her British bond holders in the autumn of 1837, and thus evoked a sharp protest from Baring Brothers on behalf of their numerous clients.[44] Even the ill-starred Chesapeake and Ohio Company, which was unable to negotiate permanent sales of its $3-million block of Maryland bonds, weathered the storm by wholesale hypothecations and large issues of canal scrip.[45]

In the revival that began early in 1838 and continued until the spring of 1839, American canal bonds found buyers in the London market. Funds were thereby provided to carry the volume of canal investment upward to a peak of $14 million in 1840. But even before the peak of the canal cycle was reached, the continuation of construction activity was doomed by the financial crisis of 1839 and the long depression that followed.

The Great Crisis. The forces that precipitated the financial panic of 1839 and eventually brought the second canal cycle to an end were manifest early in the spring of 1839.[46] A series of unfavorable developments generated mounting tensions in the London money market, and the British cotton spinners, twice victimized by Biddle, agreed to stop competing for the available supply of American cotton, thus causing a decline in its price.

Interest rates in the London money market rose rapidly after April, 1839,[47] and the impact of this development upon the efforts to finance the American canal construction was almost immediate. At the end of April, Thomas Dunlap, who succeeded Biddle as president of the Bank of the United States, purchased $1 million of Illinois sterling canal bonds under an agreement that contained a highly significant provision. The state was to be paid in the $10 notes of the Bank, which were to be distributed "under the direction of the . . . commissioners, on the public works of their state, in actual payment to those employed thereon." [48] In May, George Peabody, the London agent of the Chesapeake and Ohio Company, wrote President George C. Washington that

> capitalists show no disposition to purchase American securities to any extent at the present time, on any terms. This feeling has rendered hypothecation almost impossible; and for the loans I have contracted for you, falling due, I am called upon in every case . . . to offer at the rate of 9 and 10 percent for moneys for short periods.[49]

He subsequently placed the Company's Maryland bonds in the hands of Baring Brothers.

Indiana was the first casualty of the gathering storm. When the Morris Canal and Banking Company failed to make payments for bonds that it had purchased, work on the Whitewater Canal was suspended in August, 1839, and outlays on the Wabash and Erie thereafter were limited.[50] Elsewhere, however, the volume of construction activity continued to rise throughout the year despite the difficulties in raising funds. New York permitted the state bank purchasers of her bonds to retain the balances without interest and to defer payments until the funds were required. By these and other means, the state managed to expand construction operations through 1841. The Pennsylvania Legislature, which made large appropriations in the spring and summer of 1839, followed the example of Illinois by permitting banks to pay for canal bonds in their own $5 notes. Ohio, whose fund commissioners experienced difficulties in the summer of 1839, entered into an agreement with Baring Brothers in October, 1839, that gave the state the right to draw bills up to £30,000 without prior notice. Construction work on the Illinois and Michigan, the Chesapeake and Ohio, and the James River and Kanawha canals was expanded during 1839 by large issues of scrip, about which more will be said later.

Had the London money market recovered rapidly—within a year after the onset of the crisis—the second canal cycle would not have been interrupted. The domestic banks, which had suspended specie payments in September, 1839, could have provided temporary loans pending the recovery of the London market. However, the situation after September, 1839, grew progressively worse until a series of defaults on the foreign-held canal debts of Pennsylvania, Maryland, Indiana, and Illinois made it virtually impossible to obtain further loans. In fact, there was a large-scale repatriation of the gilt-edged New York issues.[51]

Even before the defaults, there was a disastrous decline in the prices of American canal bonds. The Chesapeake and Ohio Company, to cite a most unfortunate case, sold more than $4.5 million of its Maryland canal bonds in the period from 1839 through 1842 at about seventy-three percent of their par value. Ohio, which never defaulted, sold a block of its bonds in 1842 at sixty, and in the same year New York was compelled to raise the coupon rate to seven percent on a short-term issue.

As the prospects for further loans dimmed and the debts due contractors mounted,[52] the question of whether or not to suspend construction operations was widely discussed. Governor Seward of New York, speaking early in 1840, declared that

The sudden arrest of such expenditures and the discharge of probably ten thousand laborers, now employed on the public works, at a time when the circulation of money in other departments of business is so embarrassed as almost to have ceased, would extend throughout the whole community, and with fearful aggravation, the losses and sufferings that as yet have been confined in great measure to the mercantile class.[53]

Few of the other participants considered the income and employment effects of a suspension; rather, the tenor of the discussion tended to revolve about the direct costs that would be incurred in abandoning the unfinished works.

Notwithstanding the pervasive gloom, work on most of the major projects continued through 1841, when the value of construction work put in place—construction outlays plus obligations incurred—amounted to $11.6 million, or only $600,000 less than that in the peak year of 1840. The persistence of this high level of investment is largely ascribable to the forbearance of the contractors and the financial improvisations of the canal officials.

Many of the construction contractors on the state canal lines continued operations in the hope that regular payments would soon be resumed. In fact, work in Pennsylvania could not be halted until the Legislature, in June, 1842, denied relief to those contractors who refused to abandon their sections.

The financial improvisations, which were far more important than the willingness of the construction contractors to extend credit, fall into two classes. First, attempts were made in Ohio and Indiana to pay the contractors in land bonds. This scheme failed in Ohio, since the law provided that the land could not be sold for less than $2.50 an acre. One cannot easily appraise the results in Indiana because the land certificates issued there appeared to have also circulated as a currency.[54]

A more important attempt to improvise means of payment involved the issuance of non-interest-bearing evidences of debt which circulated as currency. There were several varieties which differed significantly with respect to liquidity. First, there was the scrip issued by chartered canal companies, the Chesapeake and Ohio, the Susquehanna and Tidewater, and the James River. These currencies were usually receivable at par value for canal tolls and water rents but were acceptable only at large discounts when offered in payment of other debts.

Certificates of indebtedness that circulated as currency were also issued by the states. Indiana, early in 1840, issued $1.2 million in state "treasury notes," which circulated until redeemed in 1856. In Ohio and Illinois, postdated checks, drawn on the state treasury and payable to the contractors, were circulated until such time as funds became available. $700,000 of these checks were issued by the Ohio commissioners in the summer of 1842, and by January, 1843, they were circulating at discounts of between forty and fifty percent.[55]

Of all the states, Pennsylvania made the most elaborate effort to continue the work on her improvement projects. Following the collapse of the Bank of the United States, the Legislature in May of 1841 passed the "Relief Act" over the Governor's veto. According to its terms, a $3.1-million loan was authorized to which all banks could subscribe by issuing "relief notes" in denominations of $1, $2, and $5. The continued circulation of these notes was assured by provisions of the act which made them redeemable in state bonds in amounts of not less than $100 and acceptable in payment of all debts due to the state. Finally, the participating banks were not required to redeem their own notes in specie until such time as the state retired the relief notes.

By January of 1842 nearly $1.75 million of the Pennsylvania relief notes were in circulation, principally along the canal lines then under construction. The state might have succeeded in completing her improvements by this scheme, had it not conflicted with the desire of the financial community to achieve a general resumption of specie payments at the earliest possible moment. It is significant that none of the larger Philadelphia banks agreed to participate in the relief note plan. The issue was resolved in March, 1842, after a disastrous run on the Bank of Pennsylvania in which the state had deposited funds for payment of interest on the improvement debt. As a result of this misfortune, for which hostile banking interests may have been responsible,[56] the Relief Act was repealed, and a swift resumption of specie payments was ordered. Within a month's time the relief notes were circulating at discounts so large that the Legislature resolved that they should be accepted in payment of taxes only if accompanied by an oath that they were not obtained at bargain rates.

In the spring of 1842, just before the general suspension of construction activity, the total circulation of canal currency—postdated checks, scrip, and so forth—reached a peak of nearly $5.5 million.[57] In view of the drastic contraction of the total money supply after the panic of 1839,[58] the canal currencies might have provided some relief from the monetary stringency, while at the same time sustaining construction operations. But they failed to achieve either purpose for reasons to which we have already alluded.

By the summer of 1842 canal construction everywhere came to a virtual standstill with the abandonment of operations along more than one thousand miles of line, which was later completed at cost of more than $57 millions.[59] Among the states, only Ohio succeeded in sustaining operations through the depression until the last of her canals was completed in 1845. Work along the line of the Wabash and Erie, while never formally abandoned, proceeded at a snail's pace during the difficult years of the early 1840s.

The Third Cycle: an Echo Phenomenon

Early in 1840, the officers of Baring Brothers circulated a letter to bankers on the difficulties in the market for American state bonds, which concluded with the following comment:

But if the old scheme of internal improvements in the Union is to be carried into effect on the vast scale and with the rapidity lately projected, and by the means of foreign capital, a more comprehensive guarantee than that of individual states will be required to raise so large an amount in a short time.[60]

These apprehensions—and the hint that a federal guarantee would be welcome—were well grounded, for, with the exception of New

York, not a single state with an extensive improvements program could meet the debt charges from current tax and canal revenues. During the high optimism of the middle 1830s it was blithely assumed that the works, even when partially completed, would yield ample net revenues. When these failed to materialize, the states resorted to the dangerous practice of meeting the interest charges from the proceeds of new loans. A crisis arrived when it was no longer possible to borrow or to persuade the state legislature to levy the requisite taxes. By 1844, about $60 million of the state bonds issued for purposes of internal improvement were in default.

The wave of defaults, which engulfed even the wealthy state of Pennsylvania, had a devastating effect upon the confidence of British investors, and it was not until the decade of the 1850s that the flow of long-term funds was resumed—this time into private railroad enterprises. In the interregnum, the perfidious states were frequently the target of indignant protests. William Wordsworth, one of the unfortunate bond-holders, penned a bitter sonnet, "To the Pennsylvanians," which concluded with

> All who revere the memory of Penn
> Grieve in the land on whose wild wood his name
> Was fondly grafted with a virtuous aim,
> Renounced, abandoned by degenerate Men
> For state dishonour black as ever came
> To upper air from Mammon's loathsome den.[61]

Baring Brothers and Company, in conjunction with other important investment bankers, resorted to the unpoetical but effective course of lobbying in the legislatures of the defaulting states. By 1848, these efforts, which did not stop short of bribery, produced acceptable settlements in Pennsylvania, Maryland, Indiana, and Illinois.[62]

The path toward fiscal virtue proved to be painful and provided the enemies of public improvements with an arsenal of effective arguments. As a result of their hostility, two substantial public projects, the Erie Extension Canal of Pennsylvania and the Whitewater of Indiana, were transferred gratis to chartered companies in 1842 and subsequently completed.[63] In the same year Pennsylvania, Illinois, and Maryland authorized the sale of all state-owned canal properties or stock, and, when responsible buyers could not be found, taxation became unavoidable. Maryland was compelled to levy a stiff general-property tax, which, after a struggle, began to yield substantial revenues in 1845. Pennsylvania in 1844, with an improvement debt in excess of $40 million—the largest in the country—enacted Draconian legislation. All corporate salaries in excess of $200 per year were taxed at the rate of two percent, and a levy of one percent was placed on incomes in excess of $200 that were earned in the "trades, occupations and professions."

Construction During the Third Canal Cycle. A third long canal cycle emerged when work was resumed on the abandoned projects. This cycle, which accounts for nearly thirty-one percent of the total canal investment in the ante-bellum period, is unique in several respects.

First, few difficulties were encountered in obtaining funds once the canal loans were authorized. There were, therefore, no interruptions of any consequence that can be ascribed to disturbances in the money markets.

Second, foreign capital played an insignificant role in the third cycle. Only about $3 million, or some five percent of the funds invested, were supplied by foreign creditors.[64] About $14 million, or nearly twenty-five percent of the total, were derived from taxes and surplus canal revenues. The remaining $15 million were easily raised in the domestic money market.

Third, the investment during this cycle was heavily concentrated in New York State. Expenditures for the Erie enlargement and the completion of the Black River and Genesee Valley canals accounted for $27.6 million or forty-eight percent of the national investment in the third cycle.

Work on the Chesapeake and Ohio Canal, the sole carry-over from the first cycle, was resumed in 1848 when a syndicate of domestic financiers purchased $833,000 of the Company's bonds at sixty percent of par, or $500,000. An additional $700,000 was to be supplied by the state of Virginia, the towns of the District of Columbia, and the contractors. After two contracting firms, which had accepted payment in company bonds, were bankrupted, the canal was finally completed to Cumberland—far short of the original Ohio River goal—in April, 1850. While gratified that the long ordeal had at last come to an end. Governor Lowe of Maryland injected a note of prescient pessimism when he commented upon the event, which a decade earlier would have elicited unrestrained enthusiasm, as follows:

Thus after twenty years of the most unprecedented difficulties and misfortune, which grew darker and thicker as it dragged its slow length along, this stupendous work has reached its destination. . . . The result is now before us. How far relief may come out of it, time alone can show. I shall not venture to fix the day nor the year.[65]

Elsewhere work on the unfinished canal lines proceeded smoothly. The Illinois and Michigan and the Wabash and Erie canals were placed under bondholder trusteeships and completed in 1848 and 1854, respectively, at a combined cost of about $5 million.[66] Work on the James River Canal was revived in 1849 and continued until the line was completed to Buchanan in 1856. The work stopped when assistance from the state of Virginia, which supplied the bulk of funds, could no longer be secured.[67] Pennsylvania resumed work on the North Branch Extension, which provided a link to the New York state canals, in 1846. She completed it, together with the enlargement of the coal carrying Delaware Division, at cost of $2.4 million in 1858. Shortly afterwards both works were sold to a railroad company.

After the suspension of construction operations on the New York State canals in March of 1842, the "Radical" Democrats limited construction expenditures to those repairs that were absolutely necessary to maintain navigation of the canals. In the normal course of events, what with the rising volume of traffic on the patently inadequate Erie Canal, the Spartan policy of the Radicals would soon have been attenuated if not completely reversed. But the Radicals managed to perpetuate their policies, long after they had ceased to be a dominant political force, by virtue of the influence that they wielded in the Constitutional Convention of 1846. There they were instrumental in obtaining the passage of an article that required that $1.85 million annually be paid of surplus canal revenues both for the retirement of the canal debt and the support of the general state government.

For several years an attempt was made to execute a construction program under the limitations imposed by the state constitution. But declining canal revenues and difficulties in forecasting the surpluses that would be available for construction purposes proved frustrating for all concerned. After an abortive attempt to circumvent the constitution, an amendment was approved in 1854 that once again permitted the state to borrow for purposes of canal construction. By 1860 $11.5 million of canal bonds were sold at prices well in excess of par. With an adequate supply of funds the Black River and Genesee Valley canals were completed in 1861, after a period of nearly twenty-six years. There was still work to be completed on the enlargement of the Erie, but, with the advent of the Civil War, the Legislature declared all contracts closed.

Canal Cycles and Business Cycles

Until now we have been principally concerned with the technological, political, and financial forces that shaped the successive canal cycles. But a number of interesting questions emerge when we reverse our field of inquiry and analyze the impacts of the canal cycles on general economic activity. How did canal construction affect the volume of business? Was it conducive to economic stability in the ante-bellum period?

The economic impacts of canal construction fall into three classes. First, there were the primary and secondary impacts of investment. Canal construction provided on-the-site employment, and the increases in income and expenditures that ensued doubtlessly gave rise to secondary or multiple effects. Second, monetary and fiscal impacts were manifested through changes in the reserve position of banks, interest rates, and tax levies. Finally, the opening of new canal routes affected the rate of economic growth by lowering transport cost and thus facilitated the migration of people and the transportation of commodities. Since the third class of impacts will be discussed at length in the next chapter, we shall here concentrate on the first two.

The Importance of Canal Investment. Before one can assess the direct and secondary effects of canal investment, some notion of its relative importance as an economic activity is required.

We estimate that canal construction at the peak of the second cycle in 1840 provided full or part-time employment for about 30,000 men. The nonagricultural labor force in the ten canal-building states in that year has been estimated by Richard A. Easterlin at 550,000.[68] Thus, canal construction may have accounted for five and one half percent of the total nonagricultural employment in those states. This proportion is neither insignificant nor impressively large. One should bear in mind, however, that the seriousness of the decline of employment in canal construction that followed after 1840 may have been mitigated by a return of many of the laborers to farms, especially in such sparsely settled states as Indiana and Illinois where the ratio of canal employment to total employment must have been far above the ten-state average

A second perspective may be gained by comparing canal investment with the new estimates of total construction which have been developed by Robert E. Gallman.[69] In 1839 the canal investment of $13.2 million was equivalent to 10 percent of the total value of construction. After that time the ratio declined sharply to a level of 0.4 percent in 1859. Canal investment in 1839 accounted for about 1 percent of the total value added by commodity production, a measure which encompasses agriculture, manufacturing the mining activities.

The Impacts of Canal Construction. The financial and developmental impacts of canal construction upon general economic activity appear to have been far stronger than those generated by the changes in the volume of investment and employment. This was especially true during the fateful second canal cycle. We remarked earlier that the great expansion of canal investment during the second cycle was not an isolated phenomenon but an integral part of a land and commodity-price boom. Let us now consider the role that canal construction played in these developments and the crises that ensued.

Canal construction during the second cycle involved transfers of funds that tended to intensify inflationary pressures. First, there was the transfer that occurred when the domestic banks negotiated bond sales in the London money market. The transfer of these funds, which was affected by foreign trade bills, improved the reserve position of the American banks and laid the basis for an expansion of their notes and deposits. A second set of monetary effects was generated by the transfer of funds from the urban centers to the sparsely settled regions through which the canal lines passed. Since the reserve ratios of country banks were notoriously low, their receipts of notes or drafts on the urban banks probably resulted in a far greater expansion of notes and deposits than that which occurred when the foreign loans were realized.

As canal construction activities were expanded, however, the increases in the demand for construction loans resulted in a veritable flood of bond offerings that strained the resources of the money markets. At the same time the opening of new canal lines accelerated the pace of land settlement and in so doing intensified speculative activity and the demand for loan funds.

Canal construction, which intensified the boom, was an exacerbating influence during the deep cyclical contraction of 1839–43. Beginning in 1840, the volume of canal investment declined precipitously, and the attempts to finance construction operations by unorthodox methods created uncertainty and pessimism within the business and banking communities. The imposition of income and new property taxes in the early 1840s also intensified the cyclical contraction.

While the second canal cycle clearly accentuated economic instability in the period 1834 to 1844, the effects of the first and third long swings in canal construction appear on balance to have been neutral. The first canal cycle occurred during a period of general economic stability, and canal expenditures during the third cycle were overshadowed by other components of capital formation.

Although they were the product of a series of unique conjunctures involving technology, sectional rivalries, international finance, and fiscal policies, the long cycles of canal construction can hardly be regarded as major economic phenomena. They were not systematically related to business fluctuations, and canal construction as a component to capital formation declined sharply in importance long before 1860. We must, therefore, conclude that real significance of canals in the ante-bellum period resides not in the repercussions of construction activity but rather in the developmental impact of canals as functioning avenues of transportation. It is to the latter aspect that we shall turn in the next chapter.

ESTIMATES OF CANAL INVESTMENT, 1815–60

H. Jerome Cranmer's estimates of canal investment, which appear in Chart 1 of the text and Table 2, were first presented in a paper on "Canal Investment 1815–1860," in *Trends in the American Economy in the Nineteenth Century* (Studies in Income and Wealth, XXIV; Princeton, 1960) pp. 547–64. In the present version, $6.39 million invested in the railroads of the Pennsylvania Mainline between 1829 and 1845 have been deducted, and the proportions of total canal investment obtained by direct and indirect methods have been indicated.

TABLE 2
Canal Investment in the United States, 1815–60
(in millions of dollars)

Year	Observed Investment	Estimated Investment	Total Investment
1815	a	a	a
1816	.02		.02
1817	.15		.15
1818	.64	.02	.66
1819	.70	.10	.80
1820	.79	.26	1.05
1821	1.29	.27	1.56
1822	2.36	.31	2.67
1823	2.46	.35	2.81
1824	1.94	.57	2.51
1825	1.62	1.07	2.69
1826	1.72	2.29	4.01
1827	2.36	3.30	5.56
1828	4.38	3.40	7.78

TABLE 2 (continued)
Canal Investment in the United States, 1815–60
(in millions of dollars)

Year	Observed Investment	Estimated Investment	Total Investment
1829	4.16	2.67	6.83
1830	5.62	1.73	7.35
1831	2.44	.99	3.43
1832	2.51	.96	3.47
1833	2.61	1.38	3.99
1834	1.44	1.09	2.53
1835	2.29	1.76	3.05
1836	2.68	1.43	4.31
1837	5.88	2.21	8.09
1838	9.81	2.00	11.81
1839	11.61	1.63	13.24
1840	12.53	2.56	14.19
1841	10.24	1.39	11.63
1842	2.88	.26	3.14
1843	.63	.34	.97
1844	.80	.15	.95
1845	1.48	.51	1.97
1846	.85	.99	1.84
1847	4.05	.11	4.66
1848	4.56		4.56
1849	3.43		3.43
1850	4.76	.11	4.87
1851	4.46	.28	4.74
1852	3.34	.01	3.35
1853	3.95		3.95
1854	4.19	.55	4.74
1855	4.46	.80	5.26
1856	3.52	.70	4.22
1857	3.08	.43	3.53
1858	2.35	.41	2.76
1859	1.74	.14	1.88
1860	1.16		1.16
Total	149.32	38.85	188.17

a Less than $10,000.

Observed and Estimated Investment. The column headed "observed investment" in Table 2 denotes that portion of the total that was obtained from the reports on annual construction outlays submitted by state canal commissions and company officials. "Estimated investment" denotes that portion which was estimated on the basis of limited information.

TABLE 3
Observed and Estimated Investment by Canal Cycles, 1815–60
(in millions of dollars)

	Observed Investment	Estimated Investment	Percent Observed	Total Investment
First Cycle, 1815–34 [a]	37.87	20.73	64.6	58.60
Second Cycle, 1834–44 [a]	59.67	12.50	82.7	72.17
Third Cycle, 1844–60	51.78	5.62	90.2	57.40
Total	149.32	38.85	79.4	188.17

[a] The investment in the trough years, 1834 and 1844, was in each case equally divided between the two cycles.

While the proportion of total investment that is based upon direct information varies between canal cycles, as Table 3 indicates, the overall proportion is nearly eighty percent.

Cranmer's "estimated investment" encompasses two classes of canals. Detailed descriptions of his estimating procedures and a critique by Harvey H. Segal will be found in *Trends in the American Economy in the Nineteenth Century*, pp. 565–70.

1. The first class contains those canals for which there is information on total costs. For each of these, Cranmer allocated expenditures to individual years by means of a time-expenditure pattern that he derived from a sample of canals for which the annual outlays were known. More than ninety percent of the "estimated investment" was obtained in this way.

2. Information on total costs was not available for canals of the second class. Cranmer used regression techniques to estimate the total cost of each and then allocated the cost over the construction period by the method described above.

Table 4 lists the canals that are included in Cranmer's totals, together with information on mileage and the method used to determine the annual investment. In the last column, the following symbols are used:

A indicates that the annual volume of investment was obtained from contemporary records.

B indicates that annual estimates of investment were obtained by allocating total costs.

C indicates that total costs were first estimated by regression techniques and then allocated.

TABLE 4
Canals Included in the Annual Estimates of Investment, 1815–60

State and Canal	Mileage	Investment Estimated by Method
Maine		
Cumberland and Oxford	20.5	B
New Hampshire		
Sewall's Falls	0.3	C
Massachusetts		
Pawtucket	1.7	B
Rhode Island		
Blackstone	45.0	B
Connecticut and Massachusetts		
New Haven and Northampton	78.0	B
New York		
Black River	35.5	A
Cayuga and Seneca	24.8	A
Champlain	81.0	A
Chemung	39.8	A
Chenango	168.2	A
Crooked Lake	7.7	A
Delaware and Hudson	108.0 [a]	C
Erie and feeders	365.5	A
Genesee Valley	124.8	A
Junction	11.0	A
Oneida Lake	5.3	A
Oswego	18.0	A
New Jersey		
Delaware and Raritan	66.0	B
Morris	103.0	C
Pennsylvania		
Bald Eagle and Spring Creek	25.0	A
Delaware Division	60.0	A
Erie and Branches	161.0	A
Lehigh Coal and Navigation	84.0	A

TABLE 4 (continued)

Canals Included in the Annual Estimates of Investment, 1815–60

State and Canal	Mileage	Investment Estimated by Method
Mainline:		
Eastern Division	42.9	A
Juniata Division	129.6	A
Western Division	105.0	A
Wisconisco Branch	13.0	A
Muncy	0.8	A
North Branch	163.3	A
Schuylkill Navigation	108.2	B
Susquehanna Division	39.0	A
Susquehanna and Tidewater	45.0 [b]	A
Union and branch	84.6	A
West Branch	80.0	A
Delaware		
Chesapeake and Delaware	14.0	B
Maryland		
Chesapeake and Ohio and branches	180.6 [c]	A
Virginia		
Albemarle and Chesapeake	43.9 [d]	C
Alexandria and Georgetown	7.1	B
James River and Kanawha	196.5	A
North Carolina		
Weldon	12.0	B
South Carolina		
Catawba	6.5	A
Drehr's	1.5	A
Lockhart's	2.8	A
Lorick's	1.0	A
Saluda	6.2	A
Wateree	4.0	A
Georgia		
Augusta	9.0	B
Brunswick	12.0	B
Ogeechee	16.0	B
Ohio		
Cincinnati and Whitewater	25.0	B
Hocking and branch	62.0	A
Miami and Erie and branches	284.3	A

TABLE 4 (continued)
Canals Included in the Annual Estimates of Investment, 1815–1860

State and Canal	Mileage	Investment Estimated by Method
Muskingum	75.0	A
Ohio and Erie and feeders	333.0	A
Ohio and Pennsylvania	101.0	A
Sandy and Beaver	84.0	B
Walhonding	25.0	A
Kentucky		
Louisville and Portland	2.3	B
Indiana		
Wabash and Erie	379.0	A
Whitewater	76.0	A-B e
Michigan		
St. Mary's Falls	1.0	B
Illinois		
Illinois and Michigan	102.0	A
Louisiana		
Company's	3.0	B
Harvey's	5.8	B
Orleans Bank	6.5	B
Texas		
Galveston and Brazos	8.0	B

[a] Includes 25.0 miles in Pennsylvania.
[b] Includes 15.0 miles in Maryland.
[c] Includes 1.2 miles in the District of Columbia.
[d] Includes 5.5 miles in North Carolina.
[e] The investment in the Whitewater Canal made by the state of Indiana before 1842 was estimated by method A.

PUBLIC FUNDS IN CANAL CONSTRUCTION, 1815–60

We estimate that $136.5 million or 73.4 percent of the total investment in canals was financed by governments. Our estimate may be broken down into two categories, the funds that were invested in the public canal systems built and operated by state governments and the government loans and stock subscriptions to mixed enterprises.

Public Canal Systems. This category includes the canals built by the following states: New York, Pennsylvania, Ohio, Indiana, Illinois, and Virginia (before 1855). This total was derived from Cranmer, "Canal Investment, 1815–1860," in *Trends in the American Economy in the Nineteenth Century* (Studies in Income and Wealth XXIV; Princeton,

N.J., 1960), but his total for "state" investment was adjusted so as to exclude the Pennsylvania railroad investments in the Mainline between Philadelphia and Pittsburgh.

Total investment in public canals: $114.3 million.

Mixed Enterprises. Stock subscriptions, gifts, and loans by governments to privately managed canal corporations are recorded in Table 5.

TABLE 5
Estimates of Public Aid to Canals, 1815–60

Canals Aided	Government	Direct Loans, Guaranteed Loans and Stock Subscriptions (in millions of dollars)
Chesapeake and Ohio	Maryland	6.09 [a]
	Georgetown, Alexandria, and Washington	1.58 [b]
	United States	1.00 [b]
	Virginia	.82 [b]
Alexandria Canal Company	Alexandria, Virginia	.47 [c]
James River and Kanawha River	Virginia	5.50 [c]
	Richmond, Lynchburg	.75 [c]
Delaware and Hudson	New York	.80 [c]
Susquehanna and Tidewater	Maryland	1.00 [a]
	Baltimore	.38 [a]
Chesapeake and Delaware	Delaware, Maryland, Pennsylvania, United States	.48 [d]
Union, Pennsylvania and Ohio, Schuylkill Navigation, Monongahela Navigation	Pennsylvania	.65 [e]
Ohio and Pennsylvania, Milan, Cincinnati and Whitewater	Ohio, Cincinnati	1.00 [f]
New Haven and Northampton	New Haven	.20 [g]
Winyaw, Wando Catawba, Wateree, Columbia	South Carolina	1.25 [h]
Several small river navigations and canals	North Carolina	.25 [i]
Total		22.22

Cycles of Canal Construction 215

TABLE 5 (continued)

a From Harvey H. Segal, The Mixed Enterprise System of Internal Improvements in Maryland (Unpublished MS).

b From Walter S. Sanderlin, The Great National Project: a History of the Chesapeake and Ohio Canal (Baltimore, 1946), p. 81; Carter Goodrich, Government Promotion of American Canals and Railroads, 1800–1890 (New York, 1960) p. 317.

c From Goodrich, Government Promotion of American Canals and Railroads, pp. 94–95, 99, and 56.

d From Balthasar Henry Meyer, ed., History of Transportation in the United States before 1860 (Washington, D.C., 1917) p. 219.

e Louis Hartz, Economic Policy and Democratic Thought, Pennsylvania, 1776–1860 (Cambridge, Mass., 1948), p. 83.

f Based upon unpublished research by Harvey H. Segal.

g Edward Chase Kirkland, Men, Cities and Transportation: a Study of New England (Cambridge, Mass., 1948).

h Goodrich, Government Promotion of American Canals and Railroads, p. 103.

i Ibid., p. 109, and Charles Clinton Weaver, Internal Improvements in North Carolina Previous to 1860 (Baltimore, 1903), pp. 48–75.

Public investment as a percentage of the total, by canal cycles, is shown in Table 6.

TABLE 6
Public Investment in Canals, 1815–60
(in millions of dollars)

	Public Investment	Total Investment	Public Investment as a Percent of Total
First Cycle, 1815–34	41.2	58.6	70.3
Second Cycle, 1834–44	57.3	72.2	79.4
Third Cycle, 1844–60	38.0	57.4	66.3
Total	136.5	188.2	73.4

Figures include the investment in the state systems of New York, Pennsylvania, Ohio, Indiana, Illinois, and Virginia (prior to 1835), plus the assistance to mixed enterprises shown in Table 5.

V. Canals and Economic Development

BY HARVEY H. SEGAL

In 1835 Michel Chevalier summarized his impressions of the American transportation network by remarking that

> The spectacle of a young people, executing in the short space of fifteen years, a series of works, which the most powerful States of Europe with a population three or four times as great, would have shrunk from undertaking, is in truth a noble sight. The advantages which result from these enterprises to the public prosperity are incalculable.[1]

The last sentence sets the theme of this chapter. Chevalier, perhaps the best informed analyst of United States transportation in the 1830s, shared a common and abiding faith in the "incalculable" advantages that would flow from canals, but he was utterly silent concerning the means by which those advantages were to be realized. His reticence was hardly unique. In fact, nowhere among the utterances of the vocal American proponents of internal improvements—a group which includes theorists, publicists, politicos, and administrators—is there a comprehensive statement of the incalculable advantages of canals or an analysis of the processes by which they were to be realized.

We shall here attempt to fulfill a task that the early proponents of canals have bequeathed to us by default—that of explaining how they accelerated economic growth in the ante-bellum United States. Our attack on the problem will proceed by a series of oblique thrusts. First, we shall briefly survey contemporary thought on the relationship between canals and economic development. Next, we shall examine the role of canals in the growth of the ante-bellum American economy. Finally, we shall appraise the successes and failures of the canal era in the light of modern economic analysis.

Canals and Development: Some Contemporary Views

Early speculation about the developmental impact of canals centered about the disposition of the commodity "surplus," a term made popular by Adam Smith to denote "provisions beyond what is necessary for maintaining the cultivators." [2] The problem was not one of producing surpluses but of providing widening markets for them in a country where resources were ample, but the areas of settlement were separated by vast distances and a formidable mountain barrier. If transport costs could be lowered through the construction of canals, regional surpluses could be exported at remunerative prices, and a rise in the level of national commodity production would ensue.

The possibility of raising the level of commodity production by lowering transport costs was clearly perceived by Albert Gallatin in his *Report* on roads and canals, issued in 1808. In the quotation that follows, Gallatin also anticipates the rudiments of benefit-cost analysis which we shall later employ.

The general utility of artificial roads and canals is at this time so universally admitted as hardly to require any additional proofs. It is sufficiently evident that whenever the annual expense of transportation on a certain route in its natural state exceeds the interest on the capital employed in improving the communication, and the annual expense of transportation (exclusively of the tolls) by the improved route, the difference is an annual additional income to the nation. Nor does in that case the general result vary, although the toll may not have been fixed at a rate sufficient to cover the interest on the capital laid out. They, indeed, when that happens lose; but the community is nevertheless benefited by the undertaking. The general gain is not confined to the difference between the expense of the transportation of those articles which had been formerly conveyed by that route, but many which were brought to market by other channels will find a new and more advantageous direction, and those which on account of their distance or weight could not be transported in any manner whatever will acquire a value and become a clear addition to the national wealth.[3]

De Witt Clinton's views were set forth in his classic 1815 "Memorial," which has been analyzed in Chapter I, and in an address to the New York Legislature in 1820, five years before the completion of the Erie. "Experience," he declared, "has evinced the precarious and fluctuating nature of foreign markets," which even in favorable years can absorb only a fraction of the country's agricultural surplus. Therefore, the realization of the potential surplus depended upon the development of domestic markets, for as Clinton explained:

Foreign commerce may cooperate in creating flourishing Atlantic cities; but internal trade must erect our towns on the lakes and rivers, and our inland villages; and internal trade must derive its principal aliment from the products of our agriculture and manufactures.[4]

But the canal, Clinton asserted, would do much more than facilitate internal commerce. An expansion of agricultural and manufacturing activities would follow from what he called the "reciprocal action of benign influences."

The reciprocal dependence of the great departments of productive industry is a wise dispensation of Providence to extend the sphere of human usefulness, to animate and multiply the motives for society, and to cement the fabric of human society. The successful progress of the important channels of communication now opening in this state will have a benign influence not only in producing the facility and cheapness of transportation, but also in creating markets for their consumption. Already do we perceive the establishment of villages on the borders of the great canal; and the raw materials of the husbandmen, obtained with the comparative ease and cheapness by the manufacturer, will be converted into articles of accommodation and comfort. This, in time, will establish on a solid foundation, an important interest, which will use the fruits of agriculture, as well in the fabrication of commodities, as in the sustenance of human life. And thus the reciprocal action of benign influences, the great departments of productive labor, will harmoniously cooperate in creating individual and national opulence. The carriers, buyers, and venders of commodities, will constitute an important class in the interior; and the great accession to other professions and pursuits, and the general augmentation of our population in consequence of our growing properity, will enable us to carry on a vast system of internal trade which will in a great measure supersede the necessity of foreign markets.[5]

When the Pennsylvania public works system was launched, Governor John Andrew Shulze succinctly summarized their anticipated development effects as follows:

There can be no doubt of the superiority of transportation by water. It brings the articles and produce so much nearer to market, that it gives a value to what would otherwise have rotted on the surface, or lain neglected in the bowels of the earth. It increases the value of his labour to the farmer, by lessening the charge of conveyance to market; and for the same reasons enables him to get his returns at a cheaper rate. It raises the price level, creates improvements, and by the consumption it occasions, and the mills and manufactureries erected, establishes a market at home; the best of all possible markets.[6]

His last sentence suggests an idea that we shall later develop: lower transport costs lead to increases in consumption which induce investment.

The economists of the period added very little to the analysis of the relationship between canals and economic development which the politicians espoused in the early stages of the first cycle of construction.

Daniel Raymond in the 1820 edition of his *Thoughts on Political Economy* defended the construction of canals by government from the charge that such expenditures would lead to a decline in total output because of the diversion of economic resources. "There is not a country on earth," he asserted, "that has not a large quantity of surplus labor." Governments could, therefore, utilize this labor without diminishing the "annual produce."[7] After laboring this point, Raymond proceeded to a defense of the New York canals, then under construction, which contains some rather vague suggestions as to their developmental impact.

There is every reason to believe that the New York Canals will increase the quantum of industry in the state equal to the whole amount of labour bestowed upon them, and that the product of labour in agriculture, manufacturers and commerce, will be as great as it would have been, had the canals not been built, so that in reality they will cause no drain on public wealth, even though they should be worth nothing to the state when finished; and it is even more probable that these enterprises have infused into the body-politic a degree of energy and industry, which will more than supply all the labour required to build the canals, and that there will be a greater product of labour in other branches of industry as a consequence of making them.8

Henry Vethake, who taught political economy at Dickinson College in the early 1820s, viewed the canal issue from a narrower, more practical vantage point. Early in 1825, a group of citizens in Cumberland County asked that he obtain information on the projected Susquehanna Canal section of the Pennsylvania Mainline with a view to determining "its relation to the prosperity of our county." Vethake then wrote to Matthew Carey of the Pennsylvania Society for the Promotion of Internal Improvements and asked

> 1st. In what manner has the construction of the turnpike road from Philada. to Pittsburg and of the Cumberland road to Wheeling, affected the comparative prosperity of the counties more or less remote from Philada. and Baltimore. I wish more particularly to know if it be possible to procure any where a comparative contemporary view of the prices of land for a series of years, before and after the construction of the roads, along these lines of communication.
>
> 2dly. The like information respecting the state of New York before, during and after, the construction of the Erie and Champlain canals.9

What Vethake sought—and probably never got—was information which could have permitted him to guage the impact of canals upon the market value of the most important component of wealth in the ante-bellum period, land.10

In view of the enormous risks borne in undertaking the early American canals, one is struck by the lack of a systematic analysis of their long-range economic impacts.11 This failure of the public officials and other proponents of internal improvements to indulge in more systematic and imaginative theorizing cannot be ascribed to the absence of revelant analytical concepts. Adam Smith's elucidation of his famous dictum, "that the division of labour is limited by the extent of the market," 12 to cite the most important example, was never exploited.

A plausible explanation of this reluctance to peer—or attempt to peer—far into the future was offered by Gallatin when he declared that the "utility of artificial roads and canals is so universally admitted as to hardly require any additional proofs." The literate public recognized that the canals would instantly lower transport costs, and, once their technical and financial feasibility was effectively demonstrated, elaborate speculation concerning their secondary impacts was hardly necessary.

The Role of Canals in American Economic Development, 1815–60

Our purpose in this section is to assess the role canals played in the development of the ante-bellum American economy, and to do this we must press the analysis of their impact beyond the initial reduction in transport costs and the disposition of the commodity surpluses. To accomplish this, we shall first introduce a simplified model of the United States economy in the early nineteenth century in which transport improvements lead to increased levels of economic activity and real income per head. This model will then be broadened to encompass the dynamic process of western settlement. Finally, we shall subject several hypotheses suggested by the model to empirical tests.

A Developmental Model. The Appalachian mountain barrier was the principal impediment to the rapid development of the American economy in the early decades of the nineteenth century. Until it was effectively breached, the economy was spatially and functionally compartmentalized. In the eastern seaboard sector, which was endowed with good harbors and a number of navigable rivers, economic development, as measured by the growth of real income per head, was probably proceeding at a moderate rate, stimulated by a rising volume of commodity exports to Europe and South America and the growth of manufacturing activities. But beyond the mountain barrier there was virtual stagnation as manifested by lightly populated though fertile lands, the widespread persistence of subsistence farming, a very low volume of regional exports and unfavorable terms of trade with the east.

The economic difficulties of the west, as its more reflective inhabitants realized, stemmed proximately from an inability to realize a great wealth potential through the disposition of the region's surplus. At the root of the problem was the high cost of transportation, which in turn acted to depress economic activity. As a consequence, a superabundant supply depressed agricultural prices in the west; and, without access to the eastern markets, where prices were considerably higher, there was little incentive to shift from subsistence farming to the production of cash crops.

The lack of efficient interregional transportation also inhibited economic development in the east. Without transport improvements its domestic market was severely limited—not only by the high costs of shipping products to the west but also by the low level of per capita income in the trans-Appalachian area.[13] By impeding interregional trade the mountain barrier thus imposed a low ceiling upon the growth potential of the American economy. So long as it remained unbroken the country would be divided into two rather distinct economic regions, tenuously linked by an uneconomical, triangular trade route via the Mississippi and the Atlantic coast.

How did canals accelerate the rate of economic growth? We shall assume that their construction led to a drastic reduction in transport costs and on this basis postulate economic changes that are viewed as a series of virtually linear causal sequences.

Let us begin with the impact of a sharp decline in transport costs on the economy of the western region. The construction of a successful interregional canal line makes it possible for the west to dispose of its agriculture surplus in the higher price markets of the east and eventually in the foreign markets. As a consequence the agricultural price differentials between the domestic markets begin to narrow. In addition to the monetary effects which follow from the improvement in the west's terms of trade, there is a shift from subsistence farming to the production of cash crops which has the significant effect of raising the level of per capita income and with it the demand for imports from the east.

In the east the developmental impacts of the transport innovation are initially transmitted through a widening of the market for its output. This widening of the east's market, made possible by lower transport costs and the rising level of per capita income in the west, induces a higher level of investment in the east. Real incomes in the east also rise with the influx of cheaper western agricultural products. The expansion of productive facilities raises the level of income in the east and at the same time increases the demand for western commodities, which in turn increases the western demand for eastern products. Thus, there is set into motion a cumulative process of interaction—a sequence of multiplier-accelerator effects—that raises the level of economic activity in both regions.

In concentrating upon the transport-cost function whose downward shift triggers the mechanism of cumulative change, we have systematized and expanded the suggestions of Gallatin, Clinton, and Shulze with the aid of the modern theories of trade and economic growth. But nothing has been said about the settlement of the western frontier, which was without doubt one of the crucial facts of economic life in the ante-bellum period. The dynamics of western settlement involved among other things the migration of people and capital, a movement that was stimulated, as we suggested in Chapter IV, by waves of speculation in commodities and land that alternated with the exigencies of the business cycle.[14]

By lowering transport costs and therefore widening the range of economic opportunity, the canals play a significant role in promoting western settlement. With the influx of population, there is an increase in the demand for nonimportable goods services which raises the output of the west and stimulated the growth of urban centers.[15] The exportable food surplus of the west is diminished in the initial stages of immigration, but the influx of capital sustains a high demand for eastern imports.[16]

The developmental forces that we have described are further reinforced by rising land prices. Because of the importance of land as a tangible asset a sharp appreciation in its value, such as that which follows from transport improvements, produces far reaching impacts. Larger mortgage loans may now be obtained by the numerous land speculators, and these lead to an increase in the money supply or a rise in the velocity of circulation, or both. While inflationary movements such as these are hardly conducive to economic stability, they accelerate the pace of economic development by enhancing general business confidence and predisposing entrepreneurs to embark upon larger and riskier ventures.

It was convenient, as a first approximation, to assume the existence of functionally distinct regions: an agricultural west and an industrial east. This dichotomy is overdrawn, and in any case the growth process that we have outlined could hardly occur without profoundly altering the economic structures of both regions, especially that of the west.

In tracing the sequence of structural changes which canals might induce, we shall begin by considering the impacts of cutting a canal through a stretch of sparsely populated country, such as that through which the Erie passed in upstate New York or the Ohio and Erie Canal in eastern Ohio. First, there are the immediate effects of construction operations. They provide a new source of employment and money income in the backward areas. Quarrying and lumbering activities may be initiated in order to supply the materials needed for the construction of locks, aqueducts, culverts, and other structures. Members of the local labor force acquire new skills while at the same time a new corps of entrepreneurs, the construction contractors, is created. Finally, one must not neglect the relatively large transfers of construction funds from the centers of population to the smaller local banks in the vicinity of construction operations. By this process, the balance of payments of the backward area is improved, and the reserve positions of local banks are strengthened, thus increasing their ability to extend loans.

The completed canal generates a series of secondary developmental effects. If it is successful as a freight carrier, the need for handling facilities—warehousing, trans-shipping, financial and distributing operations—may provide a powerful stimulus to urban growth at the terminal points.

Between the terminal points, the canal may, as the most efficient mode of transportation in the area through which it passes, stimulate local development through its power to attract economic activities that are heavily dependent upon external transport economies. The process may first begin with a concentration of commercial farming in the vicinity of the canal. If this development is followed by significant increases in population density—as a result of migration both from within and outside of the area—a market basis for the establishment of nonagricultural activities will have been established. Villages or small towns, specializing in manufacturing operations may follow. Such developments would cause property values in the vicinity of the canal to rise faster than in other areas, and this second increase in the real value of assets may act as a stimulant to local investment or to higher levels of consumer expenditures—effects that might well sustain a cumulative process of change within the area.

A two region schema, involving an east and a west, was adopted in order to emphasize the importance of interregional trade. But our analysis of the developmental role of canals would be deficient if it ended at this point. Of the four interregional canal projects, initially designed to link the eastern seaboard with the trans-Appalachian west, only one, the Erie, was successful.[17] Most of the other canals functioned as intraregional carriers or as feeders for the interregional lines. In these roles they exerted influences which were complimentary to those attributed to the interregional lines. By lowering transport costs they hastened the progress of regional specialization, raised regional output, and were conducive to urban development. Their developmental impacts, which we shall later examine in greater detail, served to reinforce the cumulative process of mutually interacting forces, which we ascribed to the expansion of interregional trade.

Measuring the Developmental Impact of Canals

We have now reached the crucial stage of our argument where it becomes necessary to clothe the bare bones of our analytical schema with more solid substance. But before we undertake this task some general warnings must be sounded.

In this classic study of the British economy, Clapham declared that "there is no way of measuring the economic gain to Great Britain from the canal system." [18] He offered no explanation for his categorical assertion, and for those such as ourselves, who have grappled with the difficulties of measurement, none is necessary. We shall, nevertheless, enter the zone that he proscribed, armed with what we hope is an adequate appreciation of the hazards involved.

The essential problem of verifying the developmental effects of canals involves the attempt to isolate causal relationships among a large number of interrelated economic variables. The difficulties involved may be illustrated by a homely parable. Once there was a man who developed an inordinate appetite for salted almonds and then regularly slaked his thirst by consuming large quantities of ice cream. Which food soon "caused" him to become obese? Was it the almonds or the ice cream? One can, of course, manage to assign a causal role to each food but not without recourse to mechanical procedures or arbitrary assumptions.

Illustrations of what we shall call the fallacy of direct imputation abound in the work of Noble E. Whitford, a state engineer and the author of an excellent history of the New York canal system. Two examples from Whitford's elaborate attempts to measure the developmental impact of the Erie Canal will suffice to underscore our point.

In his admirable collection of statistical data, Whitford presents a time-series chart that indicates that there was a sharp acceleration in the growth of New York City's population beginning in 1820 when the Erie Canal was partially completed. This intimation of a direct causal relationship between the rate of growth of the city's population and the canal is reinforced by an appeal to higher authority, for under the time-series curve Whitford placed the following quotation: "The opening of the Erie Canal gave an extraordinary impetus to the development of the city (New York), Theodore Roosevelt." [19] Elsewhere Whitford qualifies his enthusiasm when he observes that the "glorious epoch in the annals of the City" began with "the building of the Erie Canal . . . accompanied by the increase of commerce . . . the influx of aliens . . . the spread of democracy." [20] But the mischief perpetrated by the graph, Theodore Roosevelt, and Whitford's own unguarded statements is never really undone.

Another illustration is provided by Whitford's assertion that on the completion of the Erie and the Ohio canals, Cincinnati commodity prices rose "invigorated by the extension of the market and the consequent increase of demand." [21] Whitford is here dealing with one aspect of the narrowing of regional price differentials that we postulated in our theoretical model. When we examine more reliable data than Whitford's, those prepared by T. S. Berry, we discover that there was in fact a marked narrowing in the averages of western commodity price differentials between Cincinnati and New York and Cincinnati and Philadelphia in the period between 1815 and 1845.[22] But from this evidence we cannot blithely infer that the opening of the canals "caused" a decrease in supply that raised Cincinnati prices and increases in supply that lowered New York and Philadelphia prices. Among other interrelated causal factors was the migration of population. Cincinnati's population grew at a far more rapid rate between 1815 and 1845 than that of either Philadelphia or New York. The relative rise in Cincinnati prices way well have resulted from an increase in demand as well as a simultaneous decrease in supply.[23] Thus, the behavior of Cincinnati prices is consistant with the hypothesis that there were two interrelated causal factors at work—canals and migration. As in the case of our obese man, it is futile to attempt to assign them distinct roles.

Transport Costs, Commodity Trade, and Western Settlement. The decline in overland transport costs that resulted from the construction of canals can be more readily documented than any other of their developmental effects. Between 1800 and 1819, the ton-mile rate for wagon transportation, the most direct competitor of the canals, varied between 30 and 70 cents[24] and constituted a formidable barrier to the long-distance shipment of low-valued agricultural commodities. In 1817, the average rate for freight shipments between Buffalo and New York, via overland wagon and the Hudson River, was 19.12 cents per ton-mile. After the completion of the Erie Canal, in the period 1830–50, the average freight rate between Buffalo and New York declined by more than ninety-one percent to only 1.68 cents per ton-mile. By the middle 1850s, freight was moving in both directions between Albany at Buffalo at an average ton-mile cost of 1 cent, and rates on the other major canals did not exceed 2 cents.[25] In their impact upon transport costs the canals proved to be a far more radical innovation than the railroads which superseded them.[26] But these stunning results should not be exaggerated. Most of the canals were closed during winter months. Low speeds and the high costs of transferring cargoes to and from wagons at the terminals hampered their ability to compete against the turnpikes on short hauls.[27] Notwithstanding these limitations, there can be little doubt that the canals played a major role in sharply reducing the general level of overland transport costs. It is reasonable, for example, to assume that canal competition played an important role in driving the average wagon rate down to 15 cents per ton-mile by the 1850s.[28]

In the analytical model, the expansion of the interregional trade was treated as a crucial factor in the sequence of developmental impacts that followed from the construction of canals. We shall now assemble and analyze the fragmentary evidence that supports our thesis.

Of the four major canals that were projected to connect the eastern seaboard with the Ohio River, only the Erie proved successful. The Chesapeake and Ohio reached Cumberland, far short of Wheeling on the Ohio River, only in 1850; and the James River and Kanawha Canal was never extended much beyond Buchanan, a point nearly 300 miles from the Ohio. Both canals functioned essentially as intraregional carriers, a role in which they were hardly successful, as we shall later show. Some of the difficulties encountered in operating the Pennsylvania Mainline of alternating railroad and canal sections have already been described in Chapter II. Although it carried a moderately heavy volume of traffic after its completion in 1834, the Mainline failed to develop a large volume of through traffic. Total tonnage on the Allegheny Portage Railroad, which provides the best measure of through traffic that we have, never exceeded 100,000 tons per year.[29]

We turn now to the striking success of the Erie Canal as the most important interregional carrier in the overland commerce of the ante-bellum period. Its performance in this role can best be gauged by analyzing the growth of tonnage from the western states reaching tidewater.[30] This traffic, which consisted principally of agricultural commodities, grew from 54,000 tons in 1836 to a peak of nearly 1.9 million tons in 1860, a thirty-seven-fold increase. The significance of this rapid increase and the enormous expansion of the volume of western exports on which it was based emerge more clearly when it is compared with the growth of national commodity output. In Table 7 Robert E. Gallman's

TABLE 7

Growth of National Commodity Output at Constant Prices and Traffic from the West on the Erie Canal, 1839–59

Year	Commodity Output (billions of dollars) [a]	Freight from the West [b] (thousands of tons)
1839	1.09	126.7
1844	1.37	308.0
1849	1.66	768.7
1854	2.32	1,110.5
1859	2.69	1,036.6

PERCENTAGE INCREASES

Period	Commodity Output	Freight from the West
1839–44	25.5	143.1
1844–49	20.6	149.6
1849–54	39.8	44.6
1854–59	15.9	−6.7
1839–54	111.8	776.5
1839–59	145.5	718.1

SOURCES: *Historical Statistics of the United States: Colonial Times to 1957* (Washington, D.C., 1960), p. 139; "Annual Report of the Auditor of the Canal Department of the Tolls, Trade and Tonnage of the Canals of New York, February 1861," *New York Assembly Documents*, 1861, No. 93, pp. 190–91.

[a] Total values in prices of 1879.
[b] 2,000 pound tons reaching Albany from the western states.

new estimates of the value of commodity output [31]—or the value added by agricultural, mining, manufacturing, and construction activities, expressed in constant prices—and the Erie tonnage from western states are compared. Until 1849, after which railroad competition became strong,[32] shipments from the western states over the Erie increased far more rapidly than the national output.

The growth of tonnage from the west was a significant reflection of the rapid pace of settlement in Ohio, Indiana, Illinois, and Michigan. Their combined population grew from 792,000 in 1820 to more than 4.2 million in 1850, and their share of the national population soared from .8 to 18.1 percent.[33] Most prominently affected by the massive migration of population were the portions of the four states lying close to the Great Lakes along the northeastern trade route to the seaboard,[34] which were opened up by the Erie Canal. We should, of course, be guilty of the fallacy of direct imputation if we ascribed the development solely to the opening of the Erie, the Ohio and Erie, and the Wabash and Erie canals. But there is evidence enough on which to base a strong case that canals, together with the favorable trend of agricultural commodity prices and the influx of foreign capital, played an important role in the settlement of this section of the old northwest.[35] The boom in Ohio land sales in the 1820s and that which took place in Indiana between 1845 and 1852 coincided with the opening of the Erie and the Wabash and Erie canals.[36]

Moreover, the canals played a vital role in both extending the geographic scope of western trade with the eastern seaboard and altering the direction of its flow. With the completion of the Ohio and Erie from Cleveland to Portsmouth in 1830 there was a direct water route from New York to the Ohio River. In 1833 nearly 11,000 tons of eastern merchandise and salt were shipped southward from Cleveland, and by 1839 the total had risen to 30,000 tons.[37] With the opening of a large segment of the Wabash and Erie Canal in 1842—from Toledo to a point west of Fort Wayne—and with the completion of the Miami and Erie Canal, connecting Toledo and Cincinnati, in 1845, northern Indiana and western Ohio were brought into the New York trading orbit. In 1848, when the Illinois and Michigan Canal was completed, the line was extended southwestward as far as St. Louis. By 1851 more than $21 million in merchandise "imports" arrived at Chicago, largely for transshipment to markets in the southwest.[38] As a result of the widening of the market which it served, the value of the merchandise shipped to the western states on the Erie Canal rose from less than $10 million in 1836 to a peak of $94 million in 1853.[39]

Significant also was the change in the pattern of western trade that followed from the development of the northeastern route via the Great Lakes and the Erie Canal. Prior to the construction of the Erie, all save an insignificant portion of western exports were shipped southward on the Ohio and Mississippi rivers for transshipment at New Orleans. Only relatively valuable merchandise could be economically transported over the Appalachians by wagon routes. With the opening of the Erie and its major tributaries, the old triangular trade pattern was broken. The redistribution of trade in favor of the northeastern, or Erie Canal, route may be traced in Table 8, which is based upon tonnage estimates of western trade assembled by Abraham H. Sadove.[40] Although absolute volume of trade on all routes increased rapidly

TABLE 8
Distribution of the Commodity Trade of the
Northwest by Shipping Routes, 1835–53

| | PERCENTAGE OF TOTAL TRADE SHIPPED BY | | | |
Year	Northeastern Route [a]	Southern Route [b]	Eastern Route [c]	Total
1835	23.7	62.2	14.1	100.0
1839	38.2	45.4	16.4	100.0
1844	44.1	43.9	12.0	100.0
1849	52.7	38.4	8.9	100.0
1853	62.2	28.9	8.9	100.0

SOURCE: Computed from tonnage data presented in Abraham H. Sadove, Transport Improvement and the Appalachian Barrier: a Case Study in Economic Innovation (Unpublished PH.D. dissertation, Harvard University, 1950), p. 197.
[a] Via Lake Erie and the Erie Canal. [b] Down the Mississippi River.
[c] Via the Pennsylvania Mainline, Pittsburgh Turnpike, or the Cumberland Road.

between 1835 and 1853,[41] the northeastern route gained while the Ohio-Mississippi river route and eastern routes by turnpike or the Pennsylvania Mainline suffered relative declines.

Unfortunately the only evidence relating to the interregional terms of trade is based solely upon prices in the Cincinnati area —an area which was more closely tied to the south than the eastern seaboard during the ante-bellum period. But since prices there were doubtless affected by the growth of trade on the northeastern and eastern routes, we shall briefly summarize the results of Thomas S. Berry's findings.[42] The ratio of Berry's price index—A for Ohio Valley farm products to his index B—for prices of "goods brought in from outside as well as domestic manufactures," rose from an index level of 57 in 1816 to an average level of 150 in the period 1850–53. While extremely cautious for many of the reasons which we have already cited, Berry suggests that the transport innovations, together with other developments, played an important part in the improvement of the region's terms of trade.[43]

The enormous expansion of trade along the great northeastern route was highly conducive to the development of urban centers. We shall not embark upon a detailed analysis of the role of the canals in stimulating urban growth. It is sufficient to note that both Buffalo and Cleveland were insignificant villages before the construction of the Erie and the Ohio and Erie canals, and that other urban centers—such as Syracuse, Utica, Rochester, Toledo, Chicago and Peoria—benefited greatly by their canal connections.

Evidences of Local Developmental Impacts. In analyzing the developmental impacts of canals in local areas as against broad regions, we turn first to income effects of construction operations. Earlier we remarked that the volume of investment in canals, viewed from a national vantage point, was not significantly large. The picture, however, is altered somewhat when we descend to the state level. In Table 9 we compare canal investment with total income arising from commodity production and distribution by states. It is fair to state at the outset that our table biases the comparison in favor of the importance of canal construction, for, whereas 1840 was the peak year in the second canal cycle, it was also a year in which general business activity and the level of income was depressed. We assume, as seems entirely reasonable, that the overwhelming proportion of canal investment in

TABLE 9
Canal Investment and State Income Generated in Commodity Production and Distribution, 1840
(in millions of dollars)

State	Canal Investment	Total Income	Canal Investment as Percentage of Total Income
New York	$ 4.67	$150.5	3.1
New Jersey	0.19	27.2	0.7
Pennsylvania	2.93	109.3	2.7
Maryland	0.59	24.1	2.5
Ohio	2.22	60.7	3.7
Indiana	1.18	24.6	4.8
Illinois	1.12	19.6	5.7
Michigan	0.18	8.4	2.1
Virginia	0.87	59.7	1.5
Georgia	0.14	36.0	0.4
Total	14.09 [a]	521.1	2.7
United States	14.31 [a]	934.9	1.5

SOURCES: H. J. Cranmer's work sheets; R. A. Easterlin, "Interregional Differences in Per Capita Income, Population and Total Income, 1840–1950," in *Trends in the American Economy in the Nineteenth Century* (Studies in Income and Wealth XXIV; Princeton, N.J., 1960), Table A-3.

[a] $0.21 million of canal investment could not be allocated to a specific state.

each state constituted factor payments—wages, salaries, and profits—that were received by its residents. Whereas the ratios of canal investment to total income in New York and Pennsylvania are not impressively high, they increase rather sharply as we move into the more sparsely settled western states of Ohio, Indiana, and Illinois. If comparable information were available on a county rather than a state basis, there is little doubt the ratios would indicate that incomes in the backwoods areas, through which all of the major canals were cut, increased substantially as a result of construction operations.

With the great increase in interregional trade, which we have already analyzed, there was a concomitant and very rapid growth of industrial activity, especially in Pennsylvania and New York.

Manufacturing employment in those states, to cite a crude measure, more than doubled between 1820 and 1840.[44] It is doubtful whether this expansion together with the growth of the northeastern seaboard cities could have occurred without large quantities of anthracite.[45]

Until 1827 most of the coal consumed in homes and factories of the Middle Atlantic and New England states was shipped northward by coastal vessels from the Richmond area or overseas from Liverpool. After 1827, Pennsylvania overshadowed all other sources of anthracite, owing largely to the success of the four important coal canals, which commenced operations in the years indicated: the Lehigh Navigation, 1821; the Schuylkill Navigation, 1826; the Delaware and Hudson Canal, 1828; and the Delaware Division Canal of the Pennsylvania public works, 1830.[46]

In 1821 only 1,300 tons of Pennsylvania anthracite were mined,[47] but thereafter production and canal traffic increased in parallel fashion until the railroads offered stiff competition in the middle 1840s.[48] The growth of the canal-borne anthracite traffic, in thousands of short tons, was as follows: *1825*, 38; *1830*, 176; *1840*, 841; *1845*, 1,031; *1850*, 1,443; *1859*, 3,015.[49] Before 1845 the coal canals carried more than eighty percent of the annual production that was shipped out of the Pennsylvania fields; their share of the traffic after that date declined rapidly to about thirty-eight percent in 1859.[50]

By making it possible to transport coal at a rate of less than 1 cent per ton-mile by the 1840s, the canals stimulated the growth of eastern cities and made a significant contribution to the development of the iron and steel industry.

Our attempts to measure the impact of canals upon the market values of land and structures, along the lines suggested by Vethake in 1825, were for the most part frustrated by the deficiencies of the data on property assessed for purposes of taxation by counties. The county assessments of the period are notorious for understating market values by wide and variable margins. Furthermore, as portions of western Pennsylvania, large parts of

Ohio, and most of Illinois and Indiana were settled during the canal era, new counties were established. The consequent growth in the total acreage subjected to taxation makes it virtually impossible to compare the increases in land values by canal and non-canal counties. We therefore confined our investigation to New York where distortions introduced by new settlement appear to be less serious than elsewhere. Between 1820 and 1846 the value of land and improvements, adjusted for changes in the general price level, in the fourteen counties bordering on the Erie Canal increased by ninety-one percent. During the same period the real value of property in the non-canal counties, excluding New York and Kings counties (Manhattan and Brooklyn), increased by only fifty-two percent while property values in the state as a whole increased by sixty-six percent.[51]

In our model we postulated structural changes in the economies of areas contiguous to successful canals, prominent among which were shifts from agricultural to nonagricultural activities. To test the validity of this hypothesis, we shall now analyze the changes that took place in the counties bordering on the Erie Canal in New York and the principal canals in Ohio. Our investigation is confined to the period 1820–40, in order to eliminate the effects of railroads. The crude measures of changes in the structures of employment are based on county census data. Since they constitute the only available information for our test, we have used them, despite strong reservations about the degree of undercoverage and the conceptual comparability between censal years.

Table 10 shows the percentage of the population engaged in agriculture, commerce, and manufacturing in the fourteen counties bordering on the Erie Canal and a group of other counties from which we have excluded New York City and Brooklyn. Several counties traversed by lateral canals are included in the second group, but the distortions introduced by the inclusion is not appreciable.[52] Our figures clearly indicate that the employment in commerce and manufacturing grew more rapidly in the Erie Canal group than in the other county category. These dif-

TABLE 10
Changes in Occupational Distribution of Employment in
Erie Canal Counties and Others, New York, 1820–40

PERCENTAGE OF TOTAL POPULATION ENGAGED IN

	Agriculture		Commerce		Manufacturing	
	1820	1840	1820	1840	1820	1840
Erie Canal Counties [a]	19.2	20.4	.4	.9	3.8	7.1
Others: excluding New York [b] and Kings [c]	20.1	22.4	.5	.7	4.1	5.5
Kings County	7.5	6.8	.7	3.7	6.4	12.9
New York County	.8	.9	2.5	3.6	7.7	13.9
State	18.0	18.8	.7	1.2	4.4	7.1

SOURCES: *United States Census*, 1820 and 1840.

[a] Albany, Cayuga, Erie, Fulton, Herkimer, Madison, Monroe, Niagara, Oneida, Ontario, Orleans, Schenectady, Seneca, and Wayne.

[b] New York City.   [c] Brooklyn.

ferentials in favor of the Erie Canal counties, which are associated with a rapid growth of population,[53] are consistent with changes postulated in our model.

We turn now to Ohio in 1820—an underdeveloped state in which the population and industrial activity was concentrated on the Ohio River, especially in the Cincinnati area. The completion of the Ohio and Erie Canal in 1830 stimulated settlement in the lake region which ultimately—though not by 1840—reorientated Ohio's trade to the northeastern route.

Our analysis of changes in the structure of Ohio employment is more comprehensive than that for New York. Table 11 contains employment data for the counties bordering on the principal canals of the state, the Ohio and Erie in the east and the Miami in the west, together with those along the Hocking and Muskingum laterals. Since the Miami Canal was not completed to Lake Erie until 1845, we have included only the three counties lying along the completed sections.

The rate of growth of employment in both commerce and manufacturing was higher in the canal than in the non-canal counties, and this result is consistent with our model. But Table 11 reflects another development that could not have been deduced from our set of basic assumptions. Note that the shift toward nonagricultural employment in the six highly developed southwestern counties of the Miami group was far more rapid than that in the eastern Ohio and Erie Canal group. The large differential in favor of the Miami group reflects the importance of Cincinnati's Ohio River location and its southern trade. This factor appears to have more than compensated for the lack of a direct water route to Lake Erie.

When we peer below the aggregates in Table 11, a significant contrast emerges. Along the Miami Canal, the bulk of the manufacturing employment, both in absolute and relative terms, is concentrated in Hamilton County or the Cincinnati area. As one moves northward the percentage of the population engaged in manufacturing in 1840 tends to diminish from 13.9 percent in Hamilton County to 1 percent in Shelby County, the northern terminus. The single break in this pattern occurs in Miami County which includes Dayton.[54] But along the Ohio and Erie Canal the greatest concentrations of manufacturing activity are along the middle of the line rather than at the terminus. The heaviest concentrations of manufacturing activity occurred midway between Lake Erie and the Ohio River, in Stark and Musk-

TABLE 11

Changes in Occupational Distribution of Employment in Canal and Non-Canal Counties, Ohio, 1820–40

	PERCENTAGE OF TOTAL POPULATION ENGAGED IN					
	Agriculture		Commerce		Manufacturing	
	1820	1840	1820	1840	1820	1840
Ohio and Erie Counties [a]	20.6	17.0	.2	.7	3.2	4.4
Miami Canal Counties [b]	16.1	11.8	.6	1.6	4.4	9.0
Hocking Counties [c]	18.5	16.0	.1	.4	1.7	4.0
Muskingum Counties [d]	18.7	16.6	.3	.6	2.6	5.1
All Canal Counties	18.3	15.1	.3	.9	3.4	5.9
Non-Canal Counties	19.7	19.5	.2	.4	3.2	3.5
State	19.1	18.5	.3	.6	3.3	4.5

SOURCES: *United States Census,* 1820 and 1840.

[a] Coshocton, Cuyahoga, Fairfield, Franklin, Licking, Muskingum, Pickaway, Pike, Ross, Scioto, Summit, Stark, and Tuscarawas.

[b] Butler, Hamilton, Miami, Montgomery, Shelby, and Warren.

[c] Athens, Fairfield, and Hocking.   [d] Morgan, Muskingum, and Washington.

ingum counties, where the percentages of the population employed in manufacturing were 7.0 and 6.1 percent respectively. In the Cleveland area, embraced by Cuyahoga County, the percentage was only 5.2.

These contrasting patterns in the location of manufacturing activity illustrate the difficulties of isolating the developmental impacts of canals. How much weight should we attach to the Ohio River location and how much to the Miami Canal in accounting for the rapid growth of manufacturing in southwestern Ohio? Given the canal, how important was availability of raw materials in creating the spotty pattern of manufacturing activity along the Ohio and Erie? Answers to these questions—if one could, in fact, arrive at answers—would involve a consideration of the entire complex of interrelated forces bearing upon local economic development.

The uneven growth of nonagricultural employment along the Ohio canals also suggests the difficulty of extending the distinction between developmental and exploitation canals that was fruitfully applied to the New Jersey archetypes in Chapter III. Both features appear to have been embodied in the principal canals of Ohio.

Appraising the Canals

In the preceding sections we have attempted to assess the role of canals within the broad context of a changing American economy, and we have presented data on changes in population, occupational distribution, and land values that illustrate the chain of developmental processes described in our model and tend to confirm its hypotheses. The issues to which we now turn are somewhat narrower in scope, because they do not take full account of cumulative changes, but are nonetheless vital to our analysis, since they provide the basis for a comparison of the direct economic benefits conferred by the canals with the costs of their construction. A part of the answer is to be found in the profit-and-loss statements of the canals themselves, but another part must be sought in the savings in transport costs accruing to the users of the canals. We shall therefore ask the following questions: How can these benefits be measured? Which canals were successful by these criteria and which were failures? How much of the $188 million invested during the ante-bellum period was wasted?

Investment Criteria. Because of the dominant role which state governments played in constructing the ante-bellum canal network, the question of appropriate investment criteria was a matter of public concern and involved persistent and sometimes bitter controversy.

For one group of protagonists the only relevant question was the adequacy of the net return on the capital invested. This view was espoused by Silas Wright in 1827 when, as the chairman of the influential Committee on Canals of the New York Senate, he insisted that every prospective canal must promise "to reimburse the Treasury for the expense of making it." [55] Wright's academic counterpart was the economist Francis Lieber, who told his class at South Carolina College:

> No enterprise, failing by its own unprofitable nature, can be, at the same time, ruinous to adventurers, yet advantageous to the community. . . . Nothing is more common than to hear that a hotel or canal has been ruinous to adventuring individuals, but that the people have reaped the advantage of it. This cannot be. The case applied to government undertakings. They cannot be advantageous to the whole (as far as productive effects are concerned), although they would be ruinous to individuals.[56]

Dissatisfaction with the financial criterion was expressed by the Ohio Board of Public Works in 1847, when it was apparent the net revenues for the canal system as a whole were far lower than anticipated. The Board declared that

> in determining the benefits derivable from these various improvements, it might not be improper to apply the rule laid down by the early friends of the canal policy, namely that the revenue acquired from tolls is not the only advantage to be taken into account.[57]

Unfortunately the proponents of the broader view were rarely explicit about what was meant by "benefits derivable."

In appraising the performance of the canals from the narrower or pecuniary vantage point, it is well to bear in mind the following. First, fixed costs as a percentage of total operating expenses were very high. Second, net revenues, during an average operating season of about seven months, were strongly affected by the vagaries of nature. If a flash flood breeched a retaining bank and interrupted navigation for several weeks, net revenues were squeezed by the loss of tolls and the rise of repair expenditures. Third, the canals, with few exceptions, were operated as public avenues of transportation, open to the boats of all comers, and did not share in the profits of freight carriage. Fourth, tolls were almost everywhere subject to legislative regulation and, unlike carriers' charges, could not be changed rapidly enough to maximize gross revenues when the demand for transport service shifted.[58]

As a consequence of these factors, only these lines that attracted relatively heavy volumes of traffic earned enough to cover total costs, including the interest on the capital invested. The number of successes, as judged by this financial criterion, is quite small. For the period as a whole, there is the Erie, which carried an average annual volume of more than 1.7 million tons between 1836 and 1861, and a group of smaller lines, which included the Schuylkill Navigation, the Lehigh Navigation, the Delaware and Hudson, the Delaware and Raritan, the Delaware Division Canal, the Chesapeake and Delaware, the Oswego, and the Champlain. The Erie earned a surplus over all costs, including interest, as early as 1826. But on the Delaware and Raritan, Cranmer has shown that the net operating revenues did not cover the charges in the invested capital until the decade of the 1850s, when traffic increased to an average level of nearly 1.5 million tons.[59]

Financial failures are clearly dominant. Net operating revenues on the canal sections of the Pennsylvania Mainline were never equivalent to the five percent on construction costs that was required to meet interest charges. The Ohio and Erie failed to meet the test for the period as a whole, although its net earnings were well in excess of five percent on construction costs during nine years of the period 1827 through 1860. Also financial failures were the Miami and Erie, the Wabash and Erie, the Illinois and Michigan, the James River, the Chesapeake and Ohio, and a host of smaller works too numerous to list.[60]

Were we to end our analysis on this note, we should be compelled to conclude that more than two thirds of the $188 million invested in canals during the ante-bellum period constituted a social waste, or a misallocation of economic resources which might have been more effectively utilized for other purposes. But this grim view fails to take into account a crucial element, namely, the direct benefits conferred by the canals in lowering transport costs.

We shall now apply the elements of modern benefit-cost analysis to the canals of the ante-bellum period.[61] The direct benefits conferred by the canals are equal to the savings that they effected in moving the nation's commerce. Benefits from these savings, which are initially realized by the shippers, are diffused throughout the economy when the prices of commodities shipped over long distances decline. According to this criterion, a successful canal is one which confers benefits greater than its costs, including operating expenditures and charges for interest and depreciation, less tolls and other receipts. In other words, its benefit-cost ratio must be equal to or greater than one.[62]

In order to apply this concept, we must first estimate the quantity of transport service rendered by the canals in terms of the ton-miles of freight moved. Unfortunately, tonnage data for the canal system as a whole are not available, and so our analysis is confined to a sample of ten heavily utilized canals that appear in Table 12. The period 1837-46 was chosen because it is one for which data are generally available and in which railroad competition was not serious. We have adopted a ten-year average in order to eliminate the sharp cyclical fluctuations in the canal tonnage. The average hauls in Table 12 are based principally upon conjectures that we believe to be conservative. Were it possible to obtain similar data for all canals, the average annual total might be as high as 400 million ton-miles.

In estimating the direct benefits conferred by the canals—or the transport savings—we encounter a problem which is some-

TABLE 12
Average Ton-Mileage on Ten Heavily Utilized Canals, 1837–46

	Average Annual Tonnage (in thousands)	Estimate Average Haul (in miles)	Average Annual Ton-Miles (in millions)
	(1)	(2)	(1 × 2)
Erie	811	175	141.9
Pennsylvania Mainline [a]	490	100	49.0
Schuylkill Navigation	584	40	23.4
Delaware and Hudson	249	70	17.4
Lehigh Navigation	344	40	13.8
Champlain	262	50	13.1
Delaware and Raritan	249	44	11.2
Oswego	235	35	8.2
Union	103	70	7.2
Chesapeake and Delaware	154	19	2.9
Total			288.1

SOURCES: The sources from which the average annual tonnage figures were computed are as follows:

Erie, Champlain and Oswego: "Annual Report of the Auditor of the Canal Department of the Tolls, Trade and Tonnage of the Canals of New York, February, 1861," *New York Assembly Documents*, 1861, No. 93, pp. 191–98.

Pennsylvania Mainline: Data for the period 1837–39 were taken from Michel Chevalier, *Histoire et description des voies des communication aux Etats Unis et des travaux d'art qui en dependent* (Paris, 1840), I, 530; for the years 1840–46, the tonnage was estimated on the basis of gross revenues.

Schuylkill, Delaware and Hudson, and Lehigh: Chester Lloyd Jones, *The Economic History of the Anthracite-Tidewater Canals* (Philadelphia, 1908), pp. 29, 35, 86, 155.

Chesapeake and Delaware, and Union: James Weston Livingood, *The Philadelphia-Baltimore Trade Rivalry, 1780–1860* (Harrisburg, Pa., 1947), pp. 98, 114.

Delaware and Raritan: H. Jerome Cranmer, The New Jersey Canals: State Policy and Private Enterprise, 1820–32 (Unpublished PH.D. dissertation, Columbia University, 1955), pp. 315–316.

[a] Canal sections only.

times ignored in the current literature. If one could reasonably assume that all the freight sent over the canals would have been shipped by wagon or, in other words, that the price elasticity of demand for canal transportation was zero, the problem of determining canal benefits could be easily solved. Under such circumstances we should multiply the average annual ton-mileage by the savings per ton-mile effected by the canals. Wagon rates in the period 1837–46 averaged about 25 cents while the canal rate was about 2 cents per ton-mile.[63] With average savings of 23 cents per ton-mile, we thus obtain, under our restrictive assumptions, an annual benefit of $66 million.

But the $66-million figure would be correct only if all ten canals functioned as purely exploitative lines to which an existing volume of freight was shifted from the wagon haulers. The unreality of this assumption, however, is apparent when we realize how much the tonnage on the Erie and other canals exceeded the previous wagon traffic. The canals, in other words, tended to function as developmental, not purely exploitative, lines that generated traffic. As the growth of new activities was fostered by the creation of external transport economies and canal rates declined with improvements in operating techniques and increases in the capacity of the barges, an increasing number of new shippers found it profitable to utilize canal transportation. The savings per ton-mile of 23 cents, then, applied only to that small portion of the traffic which would have been carried by wagons in the absence of canals. The remaining shippers would have been willing to pay rates which varied between 25 cents and 2 cents. Consequently, the savings per ton-mile on the traffic that was generated by the canals cannot be accurately estimated. We can say only that it varied between 23 cents and zero.

From these considerations it is apparent that the upper benefit limit of $66 million must be reduced by an indeterminate but large quantity. However, during the period 1837–46, the total annual costs for the entire canal system amounted to only $3.1 million,[64] and about two thirds of this was covered by tolls and other revenues. Therefore, unless one makes the implausible assumption that the average benefit to the shipper amounted to less than .36 cents per ton-mile, it would appear that the benefits conferred by the ten successful canals exceeded the costs for the entire system. See the graphical analysis in "The Demand for Canal Transport Services and Benefits Conferred" (pp. 247–48).

That the benefits conferred by the successful lines were probably far in excess of the interest on the total canal investment does not, of course, dispose of the question of canals that were failures. Had the ante-bellum builders been endowed with perfect foresight, only heavily utilized canals would have been constructed, and the direct benefits would have been maximized.

The failures of the ante-bellum period, which conferred benefits that were not equal to costs, fall into two principal classes. There were those that failed because of underutilization even prior to the advent of effective railroad competition. Into this category fall the Chenango Canal in New York; the Erie and French Creek Division, the North Branch, the West Branch and the Wisconisco in Pennsylvania; the Hocking and Walhonding in Ohio; and the Whitewater in Indiana. The total investment on these lines prior to 1845 was about $13 million.[65]

The second, more important group, includes those canals that failed largely because of railroad competition. As effective alternatives to canal transportation for many if not all of the products shipped, the railroads sharply reduced savings effected by the canals. For example, in the ten-year period 1851–60, the ton-miles of transport service provided by the group of successful canals analyzed in Table 8 was more than twice that of the earlier period. But because of the rapid development of railroad service, it is doubtful that there was an appreciable increase in the annual benefit conferred by the canals.

Among the canals completed in the railroad era, the Chesapeake and Ohio best illustrates our point. The canal was completed to Cumberland at a cost of $11 million in 1850 some eight years after the Baltimore and Ohio railroad. Its total costs amounted to $1 million annually while average revenues in the period 1851–60 were only $154 thousand. Moreover, the average traffic load on the Chesapeake and Ohio amounted to only 263,000 tons or less than a third of that carried by the Erie Canal through a smaller prism from 1837 through 1860. Through shipments of coal accounted for nearly seventy percent of the total freight. Assuming an average haul of 150 miles, we obtain an estimate of 39 million ton-miles of service rendered annually. The crucial issue now is the savings per ton-mile. During this period the Baltimore and Ohio was carrying coal at the rate of 1.50 cents per ton-mile. The cost of coal carriage on the canal could hardly have been less than .50 cents per ton-mile, since the company's toll charge for coal amounted to .25 cents per ton-mile. Therefore, if we ignore the slow speeds and other cost-raising inconveniences, the saving effected by the canal in moving an annual average of 180,000 tons of coal could not have exceeded 1 cent per ton-mile. The upper limit for the benefits conferred by the 27 million ton-miles of coal carried was, therefore, $270,000. Benefits conferred would not then be equal to cost unless the average saving on the remaining 12 million ton-miles was equal to 4.8 cents, but, in view of the proximity of the railroad, this figure is very high.[66]

A similar case is that of the Wabash and Erie, the longest canal in the country, completed in 1851 at a total cost of $8.2 million. It differed from the Chesapeake and Ohio in only one significant respect, namely, that the bondholder trustees undertook the construction of the line between Terre Haute and Evansville in 1847, a time when the threat posed by the railroads should have been recognized. During the decade of the 1850s, the annual traffic on the Wabash and Erie was only 208,000 tons and was largely concentrated on the sections east of Terre Haute.[67] However, if we deduct the $2 million invested on the line between Terre Haute and Evansville and consider only the northeastern portion of the canal, it is probable that the benefit conferred was equivalent to the cost.

In addition to the major failures, there were a number of shorter lines, completed at a cost of $11 million, that were never substantially utilized. When this sum is added to the cost of the other failures, we obtain a grand total of $37.9 million, or twenty-one percent of the total canal investment in the ante-bellum period. The details appear in Table 13.

Although the bulk of the failures resulted from unfortunate investment decisions in the railroad era, it would be rash to conclude that the canal system as a whole had become technologically obsolete by the 1850s. According to the Andrews "Report," the best available contemporary source of information, twenty-six

TABLE 13
Investment in Canals That Failed

State and Canal	Total Investment (in millions of dollars)
New York	
Black River	3.2
Chenango [a]	2.7
Genesee Valley	5.8
Pennsylvania	
Erie Canal and French Creek Division [a]	6.4
West Branch	1.7
North Branch Extension	2.0
Wisconisco [a]	.2
Maryland	
Chesapeake and Ohio	11.0
Ohio	
Hocking [a]	.6
Walhonding [a]	.9
Indiana	
Wabash and Erie (south of Terre Haute)	2.0
Whitewater [a]	1.4
Total	37.9

SOURCES: The estimate of investment in James River and Kanawha was taken from company reports, and that for the Wabash and Erie from Elbert Jay Benton, *The Wabash Trade Route in the Development of the Old Northwest* (Baltimore, 1903), p. 88. All others are based on the 10th Census, *Report on the Agencies of Transportation in the United States* (Washington, D.C., 1883) pp. 731–64.

[a] Completed before 1845.

percent of the total domestic commerce in 1852 was shipped by canal while the railroads carried less than sixteen percent.[68] While these shares probably changed, it is unlikely that the canals were surpassed by the railroads before 1861.

Conclusions

Canals were long the province of the tow-path antiquarian, the retired engineer, and the local historian. Recent work in economic history, however, pictures them as playing a much more significant role in the development of the economy, and the present study provides substantiation for this view. A fully adequate appraisal of the canal contribution requires more information than we possess and a better understanding of the cumulative process of economic change. Nevertheless, the following conclusions appear warranted.

As effective avenues of inland commerce on which as many as 9 million tons of freight were moved in 1852,[69] the canals seem clearly to have conferred upon the ante-bellum economy direct benefits that exceeded their cost by a substantial margin. Though there were a number of costly failures, representing more than a fifth of the total investment, the system taken as a whole was successful in meeting the customary tests of benefit-cost analysis.

Examination of these direct benefits, however, does not take full account of the far-reaching consequences of the canal innovation in the cumulative process of economic development. It does not fully reflect the external economies of transport that promoted the growth of commercial farming and industrial activities and established the basis for the rapid growth of cities along the banks of the canals. But the most important contribution, in our view, is the one most difficult to measure. By accelerating the growth of interregional trade between the northern Atlantic seaboard and the territory of the old northwest, the Erie Canal and its tributaries played a vital role in extending and integrating spatially separated markets. Had there been no canals, the expansion of industrial activity in the east would have been inhibited by reliance on a much narrower domestic market and by the high cost of foodstuffs produced on inferior land; and the effective development of the west, which depended on a substantial outflow of agricultural products, could not have occurred until the coming of the railroads. By connecting the two regions, the canals initiated a sequence of cumulative impacts that promoted a rapid rate of economic growth.

THE DEMAND FOR CANAL TRANSPORT SERVICES AND BENEFITS CONFERRED

The point made on page 243 of the text can be more readily grasped with the aid of Figure 2.

Had the ten successful canals, functioning as purely exploitative lines, drawn upon a preexisting demand for transport service, the price elasticity of demand, dd, would have been zero. Under these

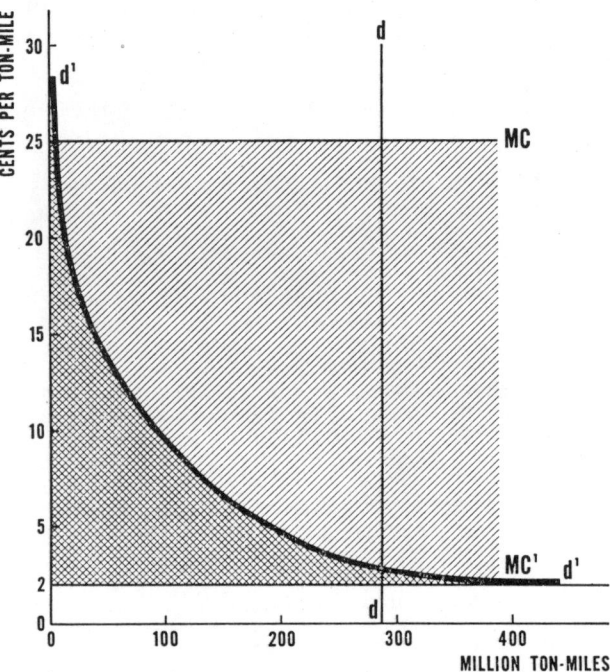

Figure 2. The Demand for Canal Transport Services and Benefits Conferred

circumstances, benefits conferred are equal to the savings per ton-mile times ton-miles of services. Benefits are represented by shaded rectangle that is bounded by dd, the ordinate and the marginal cost of wagon hauls, MC, and that for canal transportation, MC'.

The demand curve d'd' is a long-term function which encompasses the shifts in short-term functions over the ten-year period 1837–46. It is drawn on the assumption that the canals generated a demand for transport service. As consequence the shippers' surplus or benefit is drastically reduced, and there is no empirical basis for determining the relative sizes of the shaded and hatched areas. But some surplus remains unless one makes the implausible assumption of an L-shaped function bounded by the ordinate and MC'.

We assume that the cost of canal transportation was constant, but this analysis would not be substantially altered had we made a contrary assumption.

Conclusion

BY CARTER GOODRICH

Between 1815 and 1890, Americans built some four thousand miles of canals, nearly twice the mileage of the much-praised British system. Contemporary observers celebrated the achievement in glowing terms; the citizens of the young nation, it was said, had "joined the Hudson to the Mississippi . . . and changed the whole order of nature to their advantage." Chapter V has subjected claims like this to critical analysis and found in them a substantial element of truth. Although the early canals were soon supplemented and later overshadowed by the railroads, it must not be forgotten that the initial reduction in costs provided by canal transport, as compared with wagon haulage, was more drastic than any subsequent differential between railroads and canals. The effect of this reduction was decisive for the opening of substantial trade between the east and west. It was by means of canals that Americans, no longer condemned "to crawl along the outer shell" of their continent,[1] were for the first time able to make effective use of the great interior and to establish a national market on which industrial development could be based. There can be no doubt that the economic benefits conferred by the canals, though not their revenues, were considerably greater than their costs and that the system as a whole made an important contribution to the rapid growth of the American economy.

These results, however, were achieved only with difficulty and in spite of interruptions and serious errors of judgment. Canal building went forward not in a single coherent movement but, as Chapter IV indicates, in three long waves or cycles of construction. Though no regular correspondence can be established between these fluctuations and the eleven "business cycles" into which the period is commonly divided, there are clear indications of interaction between the variations in canal building and in general business activity. On the one hand, the feverish excitement of the second canal cycle was partly responsible for bringing on the panic of 1839 and the depression of the late 1830s and early 1840s, though in some cases ingenious and unorthodox methods were employed to keep the work going in the bad years. On the other hand, the effect of the business cycle on construction was most sharply felt through changes in the capital market. In a country in which, as De Tocqueville pointed out, there were almost no rich men,[2] and in which few of the states had more than rudimentary systems of general taxation, almost all of the funds for construction had to be obtained by borrowing and, in the early decades, largely from abroad. The second wave of canal building was brought to a halt by the abrupt cutting off of European funds to American borrowers; and the third wave was mainly an "echo phenomenon" devoted to the completion, largely with domestic capital, of the unfinished projects, some of which had in the meantime been rendered largely obsolescent by railroad competition.

More than a fifth of the total expenditure on canals appears from our analysis to have been devoted to canals on which the costs exceeded the benefits. Some of these seem never to have had a chance of success. Others were doomed to failure as soon as the rival innovation of the railroad, which was introduced before the end of the first cycle of canal construction, could demonstrate its practicality. Evidently, then, there were serious shortcomings in the selection, planning, and execution of canal projects. In the attempt to identify the elements of strength and weakness in the processes by which these decisions were made, the first part of the book was devoted to an analysis of the early canal programs of New York, Pennsylvania, and New Jersey.

Conclusion

The construction of canals, in these as in other cases, depended on both governmental and private decisions and was influenced by motives of public spirit as well as of individual and group interest. On the one hand, Americans often felt that it was their duty to improve the advantages given them by the Architect of the Universe, and very term "improvement" carried with it implications of enlightenment and progress similar to those now conveyed by the phrase "economic development." When De Witt Clinton, fresh from his triumph with the Erie Canal, urged New Jersey also to construct a "monument of renown" [3] for future generations, he was making an appeal that would be well understood by leaders of the underdeveloped countries of today. Advocates of the Erie and the Delaware and Raritan also made use of the argument of national defense, and the interregional canals were described as essential to the expansion and unity of the country.

On the other hand, the projects depended also on private interest and investment. As public works states, New York and Pennsylvania needed to sell their bonds in foreign markets and were able to do so in the early decades, though it became more difficult after William Wordsworth and other investors made the discovery that Pennsylvania's credit was less firm than that of the nation or of New York. As a state that obtained its improvements without public expenditures, New Jersey depended on individual subscriptions to the stock of the canal companies; and these seem to have been based almost entirely on belief in the profitability of the enterprise itself or, in the case of the Morris Canal, of the attendant banking privilege. But individual subscribers to New York's early Western Lock Navigation Company, as well as to a number of mixed enterprises in other states, seem also to have been influenced by the gains that would come to them as landowners or as shippers of produce.

Even governmental support for the canals was often urged on the grounds of the revenue they would bring in and the taxation they would render unnecessary. If only the state of New Jersey would build the Delaware and Raritan, it was argued, its college at Princeton could have the best-endowed professorships and scholarships in the world! But the proponents were often careful to add that the project would be worth undertaking even if no cent of direct return was ever received. The principal case for public support of the canals was based on the widely diffused advantages they would offer for trade and development. Farmers or prospective settlers were to gain by cheaper transport for their products, merchants by an extension of their trade, manufacturers by a wider market for their goods, mechanics and laborers by increased employment, and landowners by the enhancement of the value of their property.

In these debates little was heard of the argument that provision of transportation facilities was not a proper function of government, though the deep-rooted aversion to taxation continued to set limits on what could be undertaken. Nor was there in these early decisions any dependable correlation with the alignment of political parties. Moreover, since the benefits claimed and realized from the improvements were spread over so many occupational groups, the canal issues cannot readily be analyzed in terms of the struggle of classes. The merchant of the great cities often took a particularly prominent role, but the actual leaders in the three states under examination were surprisingly varied in origin. In the case of New Jersey's Morris Canal, leadership passed abruptly from a local patriot and promoter to the "Wall Street bees," whose interest was in the banking privilege. In the influential Pennsylvania Society, most of the membership was drawn from the merchant group; but its indefatigable leader, Mathew Carey, was a retired publisher, giving his time as a volunteer enthusiast for internal improvements. In New York, the movement for the Erie Canal made little headway while it remained in the hands of visionaries and a few western leaders and succeeded when it won the support of New York City merchants; but it required a great political leader, De Witt Clinton, to organize the victorious combination of forces.

Local rather than class interest played the greater part in the selection of projects. By their very nature, improvements in transportation, though diffusing their advantages widely among occupational groups, could hardly fail to favor certain areas more than others. Construction of a canal gave employment in one place rather than another; if it was successful, trade would follow its particular channel and not other alternatives. Spokesmen for the less favored localities were therefore likely to oppose the project on the ground that it would merely "impoverish one district for the purpose of enriching another."[4] Just as conflicts of state and regional interest had made it impossible to secure national agreement on a program like Gallatin's that attempted to keep these advantages in balance, so the decisions in these three states, as in others, turned largely on the group interests of the states themselves and local areas within them. To this we may recognize modern analogies in the efforts of a board of trade to collect contributions from business men and corporations to bring a world's fair to a particular city, in municipal expenditures on airports and sports arenas, and in the inducements offered by a number of states to attract the location of industrial establishments. It is, however, difficult to appreciate today the extent to which the sense of group interest and the emotions of patriotism and public spirit centered during the canal era on the development of individual states and cities. The competition of state against state, the rivalry of ambitious seaports like Philadelphia and New York, and the conflicting interests of local areas, played the most conspicuous part in the drama of the canal decisions. In our own three cases, the superiority of the Pennsylvania turnpikes provided one argument for the building of the Erie Canal, and New Jersey's decisions of 1824, and even more clearly Pennsylvania's decision of 1826, were largely motivated by emulation and fear of the New York achievement.

What, then, can be learned from the success and failure of the three programs? New Jersey's "improvements without public funds," as Chapter III indicates, represented neither a philosophical preference for private enterprise nor a model that could have been widely followed in other states. Construction of the "short cut" between New York and Philadelphia seems indeed to have been somewhat delayed by public and legislative reluctance to permit a private corporation to exploit such an exceptional opportunity. Construction of the more developmental Morris Canal, on the other hand, was obtained only by the grant of banking privileges at the very height of a "bank mania" that was to collapse within a year—a result due not to any Machiavellian shrewdness on the part of New Jersey legislators but rather to the overreaching eagerness of the lobbyists who influenced them.

The contrast between New York's great success with the Erie and Pennsylvania's relative failure with its Mainline is somewhat more difficult to explain. It is of course clear in retrospect that New York possessed the one best route for a canal to the west, though Chapter I points out that its selection required rejection of an easier though less magnificent alternative. It was therefore possible for the state to concentrate its initial effort mainly on the intersectional project without arousing great opposition, though at a later period the demands of "neglected" areas led to the construction of lateral canals and other improvements of more dubious value. In Pennsylvania, on the other hand, the mountains presented an obstacle that in the end could be successfully surmounted only by a railroad. An intersectional improvement therefore confronted great difficulties, and at the same time various regions of the state had their own projects, some of them entirely reasonable, for shorter canals. Under these circumstances, the advocates of the Mainline felt that they could not secure adoption of their own project without agreeing to the simultaneous construction of many miles of "branch" canals; and fear that the necessary combination of votes could not otherwise be held together influenced the decision to build a "mongrel" and inefficient system of alternate segments of canals and railroads rather than to take sufficient time to investigate the alternative of the full railroad. New York's problem was therefore the easier to solve, but this does not seem wholly to explain the contrast between the sureness of concept and execution that characterized the Erie enterprise and what Chapter II describes as the "panic reaction" of the Pennsylvania mainliners. Such striking differences in judgment and in attitude toward innovation suggest the need of going beyond the facts of geography and economics to search for other determinants of the characters of the groups concerned. Why effective leadership emerged or failed to emerge is a puzzling question deserving further analysis both in this case [5] and for many other political decisions that influenced the course of American economic development.

The particular circumstances that led American governments to promote the construction of canals in the ante-bellum period have long since passed; but the "comparative capability of nations to execute great enterprises" [6] remains a major question wherever there are important economic undertakings that cannot be left entirely in private hands. The efforts of the early American communities to provide themselves with adequate transportation facilities were in some cases characterized by bold and imaginative planning. In others they suffered from logrolling and miscalculation and similar defects of the political process, some of them peculiar to the time and others that remain difficult to avoid whenever economic decisions must be made by governmental agencies. Yet if, as we believe, the net effect of the canal movement was to give substantial impetus to economic development, it should be recalled that the Americans of the day achieved this result by using the resources of government to provide the improvements which private enterprise could not yet supply.

Notes

INTRODUCTION

1. Cadwallader D. Colden, *The Life of Robert Fulton* (New York, 1917), p. 16, referring to R. Fulton, *A Treatise on the Improvement of Canal Navigation* (London, 1796).
2. Fulton, *Treatise on the Improvement of Canal Navigation*, p. 8.
3. Baron Dupin, *The Commercial Power of Great Britain: Exhibiting a Complete View of the Public Works of This Country*, trans. from the French (2 vols.; London, 1825), I, xx, xxiii, p. xx.
4. Joseph Priestley, *Historical Account of the Navigable Rivers, Canals and Railways throughout Great Britain* (London, 1831), p. 97. This book, by an official of the Aire and Calder Navigation, gives an alphabetically arranged summary of all improvements constructed.
5. Colden, *Life of Robert Fulton*, p. 285.
6. J. H. Clapham, *An Economic History of Modern Britain*, I (Cambridge, 1926), 80–82.
7. David Stevenson, *The Principles and Practice of Canal and River Engineering* (3d ed.; Edinburgh, 1886), p. 11.
8. M. de la Lande, *Des Canaux de Navigation, et specialement du Canal de Languedoc* (Paris, 1778), p. 4.
9. Fulton, *Treatise on the Improvement of Canal Navigation*, p. 102, and H. W. Dickinson, *Robert Fulton, Engineer and Artist; His Life and Works* (London, 1913), p. 54.
10. Dupin, *The Commercial Power of Great Britain*, I, 334–35.
11. Clapham, *An Economic History of Modern Britain*, I, 75–76.
12. The Canal Commissioners of the State of New York, February 20, 1824, quoted in Noble E. Whitford, *History of the Canal System of the State of New York* (Albany, 1906), I, 794.
13. Dickinson, *Robert Fulton*, p. 37.

A British engineer writing in 1838 called attention to many of these differences and described American engineering as often representing "a judicious and ingenious arrangement to suit the circumstances of a new country in which the climate is severe—a country in which stone is scarce and wood is plentiful and where manual labor is very expensive." David Stevenson, *Sketch of the Civil Engineering in America* (London, 1838), quoted in James Kip Finch, "A Hundred Years of American Civil Engineering, 1852–1952," in American Society of Civil Engineers *Centennial Transactions*, vol. CT (1953), p. 29.

14. Daniel Hovey Calhoun, *The American Civil Engineer: Origins and Conflict* (Cambridge, Mass., 1960), esp. p. 3. On p. 52 is a tabulation of the training and background of 84 men who, in 1837, were chief engineers on canal and railroad projects. Of these 19 are listed as "Army engineers" and 11 as coming from "N. Y. canals before 1826." For the Army Engineers see Forest G. Hill, *Roads, Rails and Waterways: the Army Engineers' and Early Transportation* (Norman, Okla., 1957).

15. Albert Gallatin, *Report of the Secretary of the Treasury on the Subject of Public Roads and Canals* (Washington, D.C., 1808).

16. Carter Goodrich, *Government Promotion of Canals and Railroads, 1800–1890* (New York, 1960), Chap. II, with map and table, pp. 34–35.

17. Harvey H. Segal, Canal Cycles, 1834–1861: Public Construction Experience in New York, Pennsylvania and Ohio (Unpublished PH.D. dissertation, Columbia University, 1956).

18. [George Armroyd], *A Connected View of the Whole Internal Navigation of the United States* (Philadelphia, 1826), p. 168.

19. George Rogers Taylor, *The Transportation Revolution 1815–1860* (The Economic History of the United States, IV; New York, 1951), pp. 52–53.

20. "The short Cut for instance from New York to Phila[d] and again from Phila[d] to Baltimore. . . . those points will ear long be Laid hold of by some of our enterprizing Americans." Dickinson, *Robert Fulton,* p. 54.

I. AN INNOVATING PUBLIC IMPROVEMENT: THE ERIE CANAL

1. Lawrence Henry Gipson, *The British Empire before the American Revolution*, V (New York, 1942), 85–87; Orsamus Turner, *Pioneer History of the Holland Purchase of Western New York* (Buffalo, 1849), pp. 175–76. Colden's report, entitled "A Memorial concerning the Fur Trade of the Province of New York" is reprinted in David Hosack, *Memoir of De Witt Clinton* (New York, 1829), pp. 232–45.

2. See George Geddes, "The Erie Canal: Origins and History of the Measures That Led to Its Construction," Buffalo Historical Society *Publications*, II (1880), 265–66.

3. Projects suggested during the 1780s are described in Noble E. Whitford, *History of the Canal System of the State of New York* (Supplement to the Annual Report of the State Engineer and Surveyor of the State of New York for the Fiscal Year Ending September 30, 1905, 2 vols.; Albany, 1906), I, 19–25; and Robert Troup, *A Letter to . . . Brockhulst Livingston . . . on the Lake Canal Policy of the State of New York* (Albany, 1822), pp. 16–17 and Appendix pp. 6–7.

4. Troup, *Letter to Brockhulst Livingston*, pp. 17–22; Whitford, *History of the Canal System*, I, 28–32; and Beatrice G. Reubens, State Financing of Private Enterprise in Early New York (Unpublished PH.D. dissertation, Columbia University, 1960), pp. 10–11, 94–95.

5. Elkanah Watson, *History of the . . . Western Canals in the State of New York* (Albany, 1820), pp. 8–10.

6. *Ibid.*, pp. 19–21.

7. *Ibid.*, p. 59; Whitford, *History of the Canal System*, I, 32–33.

8. Watson, *History of the Western Canals*, pp. 19, 56–57; Winslow C. Watson, ed., *Men and Times of the Revolution; or, The Memoirs of Elkanah Watson* (2d ed.; New York, 1857), pp. 357–58.

9. Philip Schuyler to Elkanah Watson, March 4, 1792, in Watson, *Men and Times of the Revolution*, pp. 364–66.

10. Troup, *Letter to Brockhulst Livingston*, p. 14.

11. Whitford, *History of the Canal System*, I, 33–34. In 1798, the legislature incorporated the Niagara Canal Company for the construction of a canal between Lake Erie and Lake Ontario. In the following year the company reported that by its estimate tolls would cover less than half the interest on the capital required, and recommended postponement. Nothing further was done. See *ibid.*, I, 45–46.

12. Nathan Miller, "Private Enterprise in Inland Navigation: the Mohawk Route Prior to the Erie Canal," *New York History*, XXXI (October, 1950), 398–406; Whitford, *History of the Canal System*, I, 35–40, 42–43; Reubens, State Financing of Private Enterprise, pp. 103–18.

13. Miller, "Private Enterprise in Inland Navigation," pp. 401–4.

14. See the letter from Samuel L. Mitchill to David Hosack, November 15, 1828, Hosack, *Memoir of De Witt Clinton*, pp. 392–93; and Reubens, State Financing of Private Enterprise, p. 117.

15. See Miller, "Private Enterprise in Inland Navigation," p. 408.

16. *Ibid.*, p. 404.

17. Cited in Whitford, *History of the Canal System*, I, 44; see also the letter from Jonas Platt to David Hosack, May 3, 1828, Hosack, *Memoir of De Witt Clinton*, pp. 380–81.

18. See the letter from George Tibbits to Benjamin Tibbits, June 13, 1828, Hosack, *Memoir of De Witt Clinton*, p. 488.

19. Quoted in Whitford, *History of the Canal System*, I, 67.

20. See Albert Gallatin, *Report of the Secretary of the Treasury on the Subject of Public Roads and Canals* (Washington, 1808), appendix, pp. 85–86.

21. Letter of Major James Cochrane to Moses I. Cantine, February 10, 1822, cited in Whitford, *History of the Canal System*, I, 51–52.

22. Letter from Morgan Lewis to David Hosack, May 26, 1828, Hosack, *Memoir of De Witt Clinton*, p. 250.

23. Letter to John Parish, December 20, 1800, Hosack, *Memoir of De Witt Clinton*, p. 257.

24. See the letter from James Geddes to David Hosack, January 17, 1829, Hosack, *Memoir of De Witt Clinton*, pp. 265–66; and his letter to William Darby, February 22, 1822, in *Laws of the State of New York in Relation to the Erie and Champlain Canals* (2 vols.; Albany, 1825), I, 42–43. (Hereafter cited as New York *Canal Laws*.) For the lockage consideration see the essays signed "Hercules" reprinted in Hosack, p. 313; and the profile map of the Erie Canal in Whitford, *History of the Canal System*, I, facing p. 66. The early historians of the Erie bitterly disagreed about Morris's originating role. See Whitford, I, 52–54; Hosack, pp. 267–69; and Geddes, "The Erie Canal," 302–4.

25. Watson, *History of the Western Canals*, p. 66.

26. Letter from Hawley to David Hosack, July 24, 1828, Hosack, *Memoir of De Witt Clinton*, p. 301; see also the letter from Hawley's son, Elijah, April 26, 1813, in the De Witt Clinton Papers at Columbia University, V, 38. The "Hercules" papers are reprinted in Hosack, *Memoir of De Witt Clinton*, pp. 305–41. James Geddes insisted that he had previously discussed with Hawley Gouverneur Morris's suggestion of "tapping Lake Erie."

27. See Ellen E. Dickinson, "Joshua Forman, the Founder of Syracuse," *Magazine of American History*, VIII (June, 1882), 401.

28. See Joshua Forman, Letter to David Hosack, October 13, 1828, Hosack, *Memoir of De Witt Clinton*, pp. 343–46; Benjamin Wright, Letter to David Hosack, December 31, 1829, *ibid.*, p. 502; New York *Canal Laws*, I, 8–10. Despite his denial, there is evidence that Forman had read the Hercules essays before introducing his resolution. See Whitford, *History of the Canal System*, I, 57 and the references cited there.

29. Simeon De Witt to James Geddes, June 11, 1808, New York *Canal Laws*, I, 11–12.

30. Letter to William Darby, February 22, 1822, New York *Canal Laws*, I, 42–46. See also Whitford, *History of the Canal System*, I, 59–61; Geddes, "The Erie Canal," pp. 277–79; and Simeon De Witt to William Darby, February 25, 1822, New York *Canal Laws*, I, 38–42.

31. Report of James Geddes to the Surveyor-General, January 20, 1809, New York *Canal Laws,* I, 31–32.

32. Guy S. Callender, "The Early Transportation and Banking Enterprises of the States in Relation to the Growth of Corporations," *Quarterly Journal of Economics,* XVII (1903), as reprinted in Joseph T. Lambie and Richard V. Clemence, eds., *Economic Change in America* (Harrisburg, 1954), p. 525; David Maldwyn Ellis, *Landlords and Farmers in the Hudson-Mohawk Region, 1790–1850* (Ithaca, 1946), pp. 2–20, 77–82; Samuel McKee, Jr., "The Economic Pattern of Colonial New York," in Alexander C. Flick, ed., *The History of the State of New York* (10 vols.; New York, 1933), II, 263–65.

33. Cited in Geddes, "The Erie Canal," p. 300.

34. On the reasons for the trend to federal aid to internal improvements in this period see Carter Goodrich, *Government Promotion of American Canals and Railroads, 1800–1890* (New York, 1960), pp. 19–27; and William R. Willoughby, "Eary American Interest in Waterway Connections Between the East and the West," *Indiana Magazine of History,* LII (December, 1956), *passim.*

35. Gallatin, *Report,* pp. 5–7, 37–38, 41–47, 68–69, 84–86.

36. Watson, *History of the Western Canals,* pp. 59–60.

37. Article in *The Weekly Inspector,* June, 1807, reprinted in Hosack, *Memoir of De Witt Clinton,* p. 422.

38. Letter to Simeon De Witt, August, 1810, in *Remarks on the Importance of the Contemplated Grand Canal, between Lake Erie and the Hudson River* (New York (?), 1812), pp. 4–9.

39. Letter from Benjamin Wright to David Hosack, December 31, 1828, in Hosack, *Memoir of De Witt Clinton,* p. 502.

40. The resolution is reprinted in *ibid.,* pp. 344–45.

41. Whitford, *History of the Canal System,* I, 61–63; paper by Thomas Eddy, Hosack, *Memoir of De Witt Clinton,* pp. 375–77; Letter from Jonas Platt to David Hosack, May 3, 1828, *ibid.,* pp. 381–84; William A. Bird, "New York State: Early Transportation," Buffalo Historical Society *Publications,* II (1880), 28; Platt's resolution is in New York *Canal Laws,* I, 46–47.

42. See Whitford, *History of the Canal System,* I, 65. The report is reprinted in New York *Canal Laws,* I, 48–69.

43. It is curious that, despite this conviction, plans for the Champlain Canal that called for a long slack-water navigation on the upper Hudson between Fort Edward and Troy were not changed. The canal was completed in 1820, then converted to a fully independent canal during the next seven years. See Edward Chase Kirkland, *Men, Cities and Transportation: a Study of New England History, 1820–1900* (2 vols.; Cambridge, Mass., 1948), I, 85–86.

44. New York *Canal Laws,* I, 70–71.
45. Letter to David Hosack, October 13, 1828, Hosack, *Memoir of De Witt Clinton,* pp. 346–47.
46. The speech is reprinted in Hosack, *Memoir of De Witt Clinton,* pp. 359–74.
47. See Goodrich, *Government Promotion of American Canals and Railroads,* pp. 37–48; Joseph Hobson Harrison, The Internal Improvement Issue in the Politics of the Union, 1783–1825 (Unpublished PH.D. dissertation, University of Virginia, 1954), pp. 264–76; Hosack, *Memoir of De Witt Clinton,* pp. 357–59; Hugh Williamson to De Witt Clinton, March 15, 1811, Clinton Papers at Columbia University, IV, 73.
48. In a letter to M. Leray de Chaumount, Anne Cary Morris, ed., *The Diary and Letters of Gouverneur Morris* (2 vols.; New York, 1888), II, 552.
49. "Report of the Commissioners Appointed to Attend at the Seat of the General Government," including President Madison's message and the bill proposed by the commissioners, in New York *Canal Laws,* I, 91–100.
50. New York *Canal Laws,* I, 71–74, 89–90.
51. See New York *Canal Laws,* I, 64, 68, 89–90.
52. New York *Canal Laws,* I, 72, 79–87.
53. New York *Canal Laws,* I, 194–96.
54. See "The Holland Land Company and Canal Construction in Western New York: Correspondence Now First Published," Buffalo Historical Society *Publications,* XIV, 24–26.
55. See the letter from Edward P. Livingston to David Hosack, April 14, 1828, Hosack, *Memoir of De Witt Clinton,* p. 396.
56. New York State Commissioners *Report* (read in the Assembly, March 8, 1814), in New York *Canal Laws,* I, 102–4.
57. *Ibid.,* I, 101–2, 107–16.
58. See the letter from George Tibbits to Benjamin Tibbits, June 13, 1828, Hosack, *Memoir of De Witt Clinton,* pp. 488–89; William L. Stone to David Hosack, February 20, 1829, *ibid.,* pp. 430–61; *The Columbian* (New York City), December 16, 1815.
59. See New York *Canal Laws,* I, 105.
60. See Hosack, *Memoir of De Witt Clinton,* pp. 271–72. In later years, De Witt Clinton placed the entire blame for the unfortunate proposal on Gouverneur Morris, the author of the 1811 report. The commissioners, Clinton claimed, did not strike out the proposal from motives of delicacy and because of its hypothetical character. But the proposal was repeated in modified form in the 1812 report and had behind it the great prestige of the British engineer, William Weston. See Whitford, *History of the Canal System,* I, 66.

61. See Hosack, *Memoir of De Witt Clinton*, pp. 102–3; DeAlva Stanwood Alexander, *A Political History of the State of New York*, 1 (New York, 1906), p. 242.

62. See Whitford, *History of the Canal System*, I, 118.

63. Gustavus Myers, *The History of Tammany Hall* (2d ed.; New York, 1917), pp. 34–39; Dixon Ryan Fox, *The Decline of Aristocracy in the Politics of New York* (Columbia University Studies in History, Economics and Public Law, LXXXVI; New York, 1919), p. 194.

64. Alexander, *Political History of . . . New York*, pp. 182–84; Fox, *Decline of Aristocracy*, pp. 200–201.

65. See Charles G. Haines, "Introduction," in *Public Documents Relating to the New York Canals* (New York, 1821), pp. li–lii; Whitford, *History of the Canal System*, I, 72–73; Paper by Thomas Eddy, Hosack, *Memoir of De Witt Clinton*, p. 377; Alexander, *Political History of . . . New York*, pp. 244–45; Myers, *History of Tammany Hall*, pp. 30, 45–46, 73; Fox, *Decline of Aristocracy*, pp. 75–81, 148–59, 195–96. Clinton's "Memorial of the Citizens of New York, in Favour of a Canal Navigation" is reprinted in Hosack, *Memoir of De Witt Clinton*, pp. 406–21 and in New York *Canal Laws*, I, 122–41.

66. "The Holland Land Company and Canal Construction in Western New York: Correspondence Now First Published," Buffalo Historical Society *Publications*, XIV, 27–29, 63, 78, 120. The agent, Joseph Ellicott, was a leading figure in the canal movement.

67. See Joseph Austin Durrenberger, *Turnpikes: a Study of the Toll Road Movement in the Middle Atlantic States and Maryland* (Valdosta, Ga., 1931), pp. 52–58; and Catherine Elizabeth Reiser, *Pittsburgh's Commercial Development, 1800–1850* (Harrisburg, 1951), pp. 76–77.

68. See Archer Butler Hulbert, "The Erie Canal and its Rivals," *National Inland Waterways*, January, 1929, pp. 32–33.

69. See the letter of William L. Stone to David Hosack, February 20, 1829, in Hosack, *Memoir of De Witt Clinton*, pp. 431–35.

70. See William A. Bird, "New York State: Early Transportation," Buffalo Historical Society *Publications*, II (1880), 27; and Alvin F. Harlow, *Old Towpaths: the Story of the American Canal Era* (New York, 1926), p. 48. Another source of difficulty was Rochester's demand for a connection with Lake Ontario along the Genesee River.

71. Joseph Ellicott to Clinton, January 19, 1816, in the Clinton Papers at Columbia University, VI, 44; Clinton to Ellicott, February 3, 1816, in "The Holland Land Company," pp. 41–42; and Whitford, *History of the Canal System*, I, 73–74.

72. See the Commissioners' 1816 report, in New York *Canal Laws*, I, 116–18; and the letter from Ellicott to Clinton, February 25, 1816, in the Clinton Papers at Columbia University, VI, 51. The Assembly

debate is described in Geddes, "The Erie Canal," pp. 289–93. Gouverneur Morris refused to sign the report after his own draft had been rejected by the other commissioners. At issue were two of his strongly held beliefs: the inclined plane and the need for a foreign engineer. See Whitford, *History of the Canal System,* I, 74. After the 1817 decision to build the canal, Geddes and Wright dramatically demonstrated the accuracy of their work by each carrying a level in opposite directions around a circuit of almost one hundred miles. On meeting, their levels diverged by less than an inch and a half. See the Commissioners' 1818 report, in New York *Canal Laws,* I, 369–70.

73. The legislative proceedings on the bill and the act itself are in New York *Canal Laws,* I, 141–86. See also the letters of William L. Stone and James Lynch, Hosack, *Memoir of De Witt Clinton,* pp. 431–38, 462–64; Whitford, *History of the Canal System,* I, 175; and Geddes, "The Erie Canal," p. 290.

74. See New York *Canal Laws,* I, 197–98, 268–69, 301–7.

75. The offer was made in a letter to Clinton dated March 11, 1817. See the Clinton Papers at Columbia University, VII, 7; and New York *Canal Laws,* I, 285–87.

76. See the letter from Thomas Eddy to Clinton, February 15, 1817, in Clinton Papers at Columbia University, VII, 6; "The Holland Land Company," pp. 116–19; Harrison, The Internal Improvement Issue, pp. 367–70; and the letter from the New York Commissioners to members of Congress, January 22, 1817, in New York *Canal Laws,* I, 311–12. The government's bank stock amounted to $7 million; the bonus was $1.5 million. Calhoun estimated a yield from this fund of $650,000 per year. See Charles M. Wiltse, *John C. Calhoun, Nationalist, 1782–1828* (3 vols.; Indianapolis, 1944). I, 135.

77. At the time of the debate and passage of the bill, a European loan was evidently still in doubt because of strong opposition by the Dutch government to any increase of Dutch foreign lending. See Van Eeghen and Co. to William Bayard, July 21, 1817, Clinton Papers at Columbia University, VII, 29.

78. Report on the speech of Mr. Williams in the Assembly, April 9, 1817, *The Columbian* (New York City), April 16, 1817.

79. Report of the Joint Committee on Canals, read in the Assembly March 18, 1817, in New York *Canal Laws,* I, 284–85; legislative proceedings on the 1817 bill and the bill as passed, in *ibid.,* I, 335–64; *The Columbian* (New York City), April 16, 1817; Whitford, *History of the Canal System,* I, 82–86, 98; Hosack, *Memoir of De Witt Clinton,* 489–93. The 1817 bill, like that of 1816, called for construction only of the relatively easy middle section between Rome and the Seneca River. A law of April 7, 1819, authorized completion of the canal.

During the session of 1820, opponents of the canal tried unsuccessfully to postpone construction west of the Seneca in the hope that, with construction of the eastern section, the benefited areas would withdraw from the movement for completion. See letter to David Hosack from William L. Stone, February 20, 1829, Hosack, *Memoir of De Witt Clinton*, pp. 454–55; Whitford, *History of the Canal System*, I, 91–95; Harlow, *Old Towpaths*, pp. 57–58; Peter Ploughshare (pseud. of Samuel B. Beach[?]), *Considerations against Continuing the Great Canal West of the Seneca* (Utica, N.Y., 1819), *passim*.

80. See the report on the debate by William L. Stone in Hosack, *Memoir of De Witt Clinton*, pp. 451–53; and Jabez D. Hammond, *The History of Political Parties in the State of New York* (2 vols., Albany, 1842), I, 441.

81. Myers, *History of Tammany Hall*, pp. 42–43.

82. Fox, *Decline of Aristocracy*, p. 158, writes of the landed aristocracy of New York as "preaching the gospel of a generous development of the imperial resources of the state, which sounded to the ears of the mechanic and his party in the city, as vague, star-misty stuff."

83. See the letter from Jonas Platt to David Hosack, May 3, 1828, Hosack, *Memoir of De Witt Clinton*, pp. 386–88; and William Kent, *Memoirs and Letters of James Kent* (Boston, 1898), pp. 168–170.

84. See Hulbert, "The Erie Canal and its Rivals," pp. 33–34. In his *The Welland Canal Company* (Cambridge, Mass., 1954), pp. 19–20, Hugh G. J. Aitken suggests that the revival of the project after 1815 was the consequence of a sudden change of sentiment in favor of the Erie route, and that this occurred because of the recognition that, since armed aggression against Upper Canada had failed, Montreal's dominance of Lake Ontario would have to be accepted. However, the literal wording of Clinton's New York "Memorial," which Aitken cites, does not support this view; more important, the entire leadership of the state canal movement had been won over to the Erie route before the war began.

85. Letter to David Hosack, May 3, 1828, Hosack, *Memoir of De Witt Clinton*, p. 388.

86. See Israel D. Andrews, *Report . . . on the Trade and Commerce of the British North American Colonies, and upon the Trade of the Great Lakes and Rivers* (Senate Executive Document 112, 32d Congress, 1st session; Washington, 1854), p. 240; and Robert Greenhalgh Albion, *The Rise of New York Port, 1815–1860* (New York, 1939), p. 37.

87. See Albion, *Rise of New York Port, passim*; and his "New York Port and its Disappointed Rivals, 1815–1860," *Journal of Business and Economic History*, III (August, 1931), 608–9.

88. On the innovating aspects of the canal, see Goodrich, *Government Promotion of American Canals and Railroads,* pp. 53–54; Harlow, *Old Towpaths,* pp. 45, 51–53; Hulbert, "The Erie Canal and Its Rivals," p. 54. For a detailed history of the canal's construction, see Whitford, *History of the Canal System,* I, 86–118. The canal as a school of the nation's engineers is treated in Daniel Hovey Calhoun, *The American Civil Engineer: Origins and Conflicts* (Cambridge, Mass., 1960), pp. 24–27, 30, 34–37; and in Whitford, I, 786–807.

89. Hosack, *Memoir,* pp. 346–48.

II. AN IMITATIVE PUBLIC IMPROVEMENT: THE PENNSYLVANIA MAINLINE

1. Carter Goodrich, *Government Promotion of American Canals and Railroads, 1800–1890* (New York, 1960), pp. 61–63.

2. Joseph Austin Durrenberger, *Turnpikes: a Study of the Toll Road Movement in the Middle Atlantic States and Maryland* (Valdosta, Ga., 1931), pp. 126–28.

3. A. E. Parkins, "The Development of Transportation in Pennsylvania," Geographic Society of Philadelphia *Bulletin,* XIV (1916), 107–8.

4. Durrenberger, *Turnpikes,* pp. 52–58.

5. Abraham H. Sadove, Transport Improvement and the Appalachian Barrier: a Case Study in Economic Innovation (Unpublished PH.D. dissertation, Harvard University, 1950), pp. 138–39, 309; Catherine Elizabeth Reiser, *Pittsburgh's Commercial Development, 1800–1850* (Harrisburg, Pa., 1951), pp. 76–77; Louis C. Hunter, *Steamboats on the Western Waters: an Economic and Technological History* (Cambridge, 1949), pp. 17–22, 59.

6. Sadove, Transport Improvement and the Appalachian Barrier, pp. 164–66, 171–79; Hunter, *Steamboats on the Western Waters,* pp. 482–83.

7. For a description of the rival projects see Julius Rubin, Imitation by Canal or Innovation by Railroad: a Comparative Study of the Response to the Erie Canal in Boston, Philadelphia, and Baltimore (Unpublished PH.D. dissertation, Columbia University, 1959), *passim.* This dissertation is published in revised form in American Philosophical Society *Transactions,* LI, Pt. 7, under the title *Canal or Railroad? Imitation and Innovation in the Response to the Erie Canal in Philadelphia, Baltimore, and Boston.*

8. G. S. Callender, "The Early Transportation and Banking Enterprises of the States in Relation to the Growth of Corporations," *Quarterly Journal of Economics,* XVII (1903), as reprinted in Joseph

T. Lambie and Richard V. Clemence, eds., *Economic Change in America* (Harrisburg, Pa., 1954), p. 525.

9. On the settlement of New York and Pennsylvania see Sadove, *Transport Improvement*, pp. 330–32, 353, 429–30; and Beatrice G. Reubens, State Financing of Private Enterprise in Early New York (Unpublished PH.D. dissertation, Columbia University, 1960), pp. 16–18. On the differences in social structure see Solon J. and Elizabeth H. Buck, *The Planting of Civilization in Western Pennsylvania* (Pittsburgh, 1939), p. 129; Arthur M. Schlesinger, *The Colonial Merchants and the American Revolution, 1763–1776* (New York, 1918), p. 27; Thomas C. Cochran, *New York in the Confederation: an Economic Study* (Philadelphia, 1932), p. 182; Jabez D. Hammond, *The History of Political Parties in the State of New-York* (2 vols.; Albany, 1842), I, 32. On sectional interests in Pennsylvania see Avard L. Bishop, "The State Works of Pennsylvania," *Connecticut Academy of Arts and Sciences Transactions*, XIII (November, 1907), pp. 184–85.

10. On the Society's history and activities, see Rubin, *Imitation by Canal or Innovation by Railroad*, pp. 57–61 and the references cited there. The quotation is from the unsigned preface to William Strickland, *Reports on Canals, Railways, Roads, and Other Subjects, Made to the "Pennsylvania Society"* (Philadelphia, 1826).

11. U.S. House, Roads and Canals Committee, *Report . . . on the Chesapeake and Ohio Canal, April 17, 1834*, House Report 414, 23d Cong., 1st Sess. (Washington, 1834), pp. 3–4; George Washington Ward, *The Early Development of the Chesapeake and Ohio Canal Project* (Baltimore, 1899), pp. 44–45, 68; Walter S. Sanderlin, *The Great National Project: a History of the Chesapeake and Ohio Canal Company* (Baltimore, 1946), p. 51.

12. Bishop, "State Works of Pennsylvania," pp. 171–73; *A Compilation of the Canal and Railroad Laws of Pennsylvania* (Harrisburg, Pa., 1836), pp. 3–4.

13. Pennsylvania Board of Commissioners for . . . Internal Improvement, *Report*, February 2, 1825, Pennsylvania House *Journal*, 1824/25, Vol. II, *passim*.

14. Pennsylvania Board of Commissioners for . . . Internal Improvement, *Report of Charles Trcziyulney*, February 21, 1825, Pennsylvania House *Journal*, 1824/25, Vol. II, *passim*.

15. Only one of the three possible routes was seriously explored. See the report of the meeting in Centre County in the Philadelphia *United States Gazette*, February 14, 1825.

16. Thurman Van Metre, *Early Opposition to the Steam Railroad* (New York, 1924?), pp. 15–19; Lewis H. Haney, "A Congressional History of Railways in the United States to 1850," University of Wisconsin

Bulletin No. 211 (Madison, Wis., 1908), p. 23; John Stevens, *Documents Tending to Prove the Superior Advantages of Rail-Ways* (New York, 1812), reprinted in *The Magazine of History*, XIV, Extra No. 54 (1917), *passim*.

17. Historical Records Survey, New Jersey, *Calendar of the Stevens Family Papers: Preliminary Volume* (Newark, N.J., 1940), pp. 75–76, letters dated December 4, 1820, January 5 and 15, 1821.

18. James Weston Livingood, *The Philadelphia-Baltimore Trade Rivalry, 1780–1860* (Harrisburg, Pa., 1947), p. 142.

19. *Ibid.*, pp. 142–44; Dorothy Gregg, The Exploitation of the Steamboat: the Case of Colonel John Stevens (Unpublished PH.D. dissertation, Columbia, 1951), pp. 313–16; Archibald Douglas Turnbull, *John Stevens* (New York, 1928), pp. 447–48, 467; Van Metre, *Early Opposition to the Steam Railroad*, p. 19; Seymour Dunbar, *A History of Travel in America* (4 vols.; Indianapolis, 1915), III, 892.

20. Dunbar, *History of Travel in America*, III, 894.

21. W. T. Jackman, *The Development of Transportation in Modern England* (2 vols.; Cambridge, 1916), II, 477–83, 510n, 531–32, 535–45; George S. Veitch, *The Struggle for the Liverpool and Manchester Railroad* (Liverpool, 1930), p. 33; Strickland, *Reports on Canals, Railways, Roads*, p. 23.

22. *Poulson's American Daily Advertiser*, January 25, February 3, February 9, 1825. See also the issue of March 11, 1825.

23. *United States Gazette*, February 11, 1825; Dunbar, *History of Travel in America*, III, 178n; William Bender Wilson, *History of the Pennsylvania Railroad Company* (2 vols.; Philadelphia, 1899), I, 96–98. I have not been able to find any report by this Senate committee.

24. Pennsylvania Society for the Promotion of Internal Improvement, *Railways*, (2d ed.; Philadelphia, 1825), pp. 1–2.

25. "Tweed," *United States Gazette*, August 30, 1825.

26. *Internal Improvement. Rail Roads, Canals, Bridges, etc.* (Philadelphia, 1825), pp. 21, 25–27. According to Agnes Addison Gilchrist, *William Strickland, Architect and Engineer, 1788–1854* (Philadelphia, 1950), p. 67, this pamphlet contains the first printed suggestion for a transcontinental railroad.

27. "Canals and Railways—No. IV," *United States Gazette*, February 16, 1825, signed "Fulton." On the question of authorship, see Kenneth Wyer Rowe, *Mathew Carey: a Study in American Economic Development* (Baltimore, 1933), pp. 89–93 and bibliography.

28. See Richard I. Shelling, "Philadelphia and the Agitation in 1825 for the Pennsylvania Canal," *Pennsylvania Magazine of History and Biography*, LXII (April, 1938), 181.

29. Stevens was trying to obtain a New Jersey railroad charter and

to revive his charter for a railroad from Philadelphia to the Susquehanna. See Historical Records Survey, New Jersey, *Calendar of the Stevens Family Papers: Preliminary Volume*, pp. 81–83, Documents 5417 and 5461.

30. Pennsylvania Society for the Promotion of Internal Improvement, Acting Committee, *First Annual Report* (Philadelphia, 1826), p. 34, "Mr. Strickland's Instructions (copy)." Strickland left for Europe on March 20.

31. Philadelphia *National Gazette*, March 24, 1825.

32. *United States Gazette*, March 22, 1825. The article, dated March 4, is signed by the fifteen members of the Acting Committee, headed by Mathew Carey.

33. "No. VI. Internal Improvement," Harrisburg, March 28, 1825, signed "Fulton," pp. 4–5.

34. *A Compilation of the Canal and Railroad Laws of Pennsylvania*, pp. 4–7.

35. Bishop, "State Works of Pennsylvania," p. 180.

36. Shelling, "Philadelphia and the Agitation in 1825 for the Pennsylvania Canal," p. 185.

37. Bishop, "State Works of Pennsylvania," p. 266. *Niles' Weekly Register*, XXIX (September 24, 1825), 63, quotes the *Pennsylvania Intelligencer* as to the occupational distribution at the convention. Among the 117 delegates were 47 lawyers, 30 farmers, 16 merchants, 7 manufacturers, and 2 mechanics; the remaining 15 were miscellaneous businessmen and professionals.

38. *Niles' Weekly Register*, XXIX (September 24, 1825), 62; Shelling, "Philadelphia and the Agitation in 1825 for the Pennsylvania Canal," p. 197. See also William M. Meigs, *The Life of Charles Jared Ingersoll* (Philadelphia, 1897), pp. 153–54.

39. 3d ed.; Philadelphia, 1825. Henceforth to be referred to as *Facts and Arguments*. The pamphlet has been mistakenly attributed to Mathew Carey and to William Strickland. The author was probably George Washington Smith. For the evidence on authorship, see Rubin, Imitation by Canal or Innovation by Railroad, pp. 83–84, note 2.

40. *Internal Improvements—Railroads and Canals, No. II*, p. 3. Signed "Hamilton." See note 53, below.

41. *Facts and Arguments*, pp. 22–25, 52, 58–59.

42. *United States Gazette*, July 11, 1825.

43. *Ibid.*, July 26, 1825.

44. It was dated Edinburgh, June 5. Both the *United States Gazette* and *Poulson's* carried it.

45. *Niles' Weekly Register,* XXIX (September 17, 1825), 35. There are no further reports of this project.
46. *United States Gazette,* October 14, 1825.
47. *Ibid.,* November 23, 1825. The latter report was undoubtedly false.
48. *Ibid.,* December 29, 1825, reprinted from the London *John Bull.*
49. Pennsylvania Society for the Promotion of Internal Improvement, Acting Committee, *First Annual Report,* (January, 1826), pp. 19-20.
50. Pennsylvania Society for the Promotion of Internal Improvement, Acting Committee, *Internal Improvement. Extracts from Correspondence with William Strickland* (Philadelphia, 1825).
51. See especially, Pennsylvania Society for the Promotion of Internal Improvement, Acting Committee, *First Annual Report,* p. 9.
52. See the *United States Gazette's* correction of its error in its September 28, 1825, issue; the article by "H" in *ibid.,* September 5, 1825, and by "S" in *ibid.,* September 6, 1825.
53. There were two series of pamphlets: *Internal Improvement—Nos. I–VI. Railways and Canals,* all signed "Hamilton," and dated September through December 1825; and *Internal Improvement. New Series, I and II,* both signed "Fulton," and dated January 19 and 31, 1826. In addition, a number of short responses to opponents appeared in the *United States Gazette.*
54. He had retired in 1824 with an assured annual income of about $8,000. See Rowe, *Mathew Carey,* p. 23.
55. "Alnus," *United States Gazette,* November 8, 1825, referring to the man who signed himself "Hamilton."
56. For a more detailed discussion of the technical aspects of the debate, see Rubin, Imitation by Canal or Innovation by Railroad, pp. 109-48, and the references cited there.
57. Article reprinted from *The Scotsman* in the *American Mechanics Magazine,* I (April 9, 16, 23, 1825), and in the Philadelphia *National Gazette,* February 26, 1825.
58. Reprinted from the Boston *Palladium* in *U.S. Gazette,* March 23, 1825.
59. "Alnus," *United States Gazette,* December 30, 1825.
60. George Edward Reed, ed., *Pennsylvania Archives, Fourth Series* (Harrisburg, Pa., 1900), V, 590-93.
61. Board of Canal Commissioners, *Report* read in the Senate on February 7, 1826 (Harrisburg, Pa., 1826), *passim;* Pennsylvania Society for the Promotion of Internal Improvement, Acting Committee, *The Union Canal. Paper read at a meeting of the Acting Committee,* August 26, 1826, p. 1.

62. *United States Gazette,* January 20, 1826; *Niles' Weekly Register,* XXIX (February 18, 1826), 406–7.

63. *Pennsylvania Archives, Fourth Series,* V, 652–57.

64. A revealing postscript was added to the Pennsylvania railroad debate in August, 1826, when the Pennsylvania Society published a collection of Strickland's reports from Europe on internal improvements and omitted the passage that came out so unqualifiedly for railroads. The circumstances of the deletion were later described by John K. Kane in his obituary notice of Strickland. Kane was a vice-president of the Society and its proofreader on the Strickland book. He was about to remit the passage to the printer when "the Society's committee, and I think the Society itself, remonstrated strenuously against so perilous a committal on the part of a gentleman, whose opinions might be confounded with their own." Kane gave in, rewrote the passage, "and so saved Strickland from declaring in advance what a large part of the world knows now to be true." See Kane, "William Strickland," in American Philosophical Society *Proceedings,* VI (1854), 30; and Strickland, *Reports on Canals, Railways, Roads.*

65. See above, pp. 86–87.

66. Quoted in *Niles' Weekly Register,* XXIX (November 11, 1825), 164.

67. March 12, 1825.

68. "Canals and Railways—No. VI," *United States Gazette,* March 28, 1825, signed "Fulton."

69. March 22, 1825.

70. July 11, 26, and 27, 1825.

71. October 18, 1825.

72. February 3, 1826.

73. "Canals vs. Railroads," *United States Gazette,* June 23, 1825.

74. *Internal Improvement—No. II. Railways and Canals,* p. 2.

75. Pittsburgh *Gazette,* July 7, 1818, quoted in Livingood, *Philadelphia-Baltimore Trade Rivalry,* p. 20.

76. *Address,* p. 12.

77. "Northumberland," *United States Gazette,* September 22, 1825.

78. Letter in *United States Gazette,* March 7, 1825.

79. *Ibid.*

80. *Brief View of the System of Internal Improvement in the State of Pennsylvania* (Philadelphia, 1831), p. 12.

81. "Northumberland," *United States Gazette,* September 22, 1825.

82. *Internal Improvement—No. I. Railroads and Canals,* pp. 1–2.

83. *United States Gazette,* November 28, 1825.

84. *Ibid.,* March 7, 1825.

85. Louis Hartz, *Economic Policy and Democratic Thought: Penn-*

sylvania, 1776–1860 (Cambridge, Mass., 1948), pp. 148–49; Bishop, "State Works of Pennsylvania," pp. 186–87, 190–95, 199–204, and map facing p. 96.

86. *To the Citizens of the Commonwealth of Pennsylvania: Internal Improvement,* Philadelphia, March 6, 1827, signed "M. C.," p. 1.

87. See *Internal Improvement,* dated November 12, 1829, signed "Hamilton," *passim;* and *Internal Improvement No. II,* dated December 5, 1829, signed "Hamilton," *passim.*

88. *Brief View of the System of Internal Improvement in the State of Pennsylvania,* pp. 16–18.

89. See pp. 14, 134–36, 292.

90. Speeches at the Harrisburg Convention by James M. Porter and N. H. Loring as reported in the *United States Gazette,* August 11 and 26, 1825.

91. Quoted in Bishop, "State Works of Pennsylvania," pp. 182–83, from the Harrisburg *Chronicle* of March 10, 1825, which had copied it from the Erie *Gazette.*

92. *A Compilation of the Canal and Railroad Laws of Pennsylvania,* pp. 27–28; Bishop, "State Works of Pennsylvania," 190–92; Livingood, *Philadelphia-Baltimore Trade Rivalry,* p. 110; *Niles' Weekly Register,* XXXII (June 28, 1827), 274.

93. On the description and history of the mainline after the 1826 decision, and of the Pennsylvania Railroad, see Rubin, Imitation by Canal or Innovation by Railroad, pp. 97–106, and references cited there.

94. George Rogers Taylor, *The Transportation Revolution, 1815–1860* (New York, 1951), p. 44.

95. Robert G. Albion, *The Rise of New York Port* (New York, 1939), p. 378. Bishop, "State Works of Pennsylvania," pp. 238–39, puts the original cost at $12,106,788.

96. William H. Dean, Jr., The Theory of Geographic Location of Economic Activities with Special Reference to Historical Change (Unpublished PH.D. dissertation, Harvard University, 1938), p. 195; and Bishop, "State Works of Pennsylvania," p. 153.

97. Bishop, "State Works of Pennsylvania," p. 280. Bishop gives the year-by-year figures for each section of the mainline on pp. 278–80. Harvey H. Segal, Canal Cycles, 1834–1861: Public Construction Experience in New York, Pennsylvania, and Ohio (Unpublished PH.D. dissertation, Columbia University, 1956), p. 214, gives net revenues as percent of cost, year by year.

98. Bishop, "State Works of Pennsylvania," pp. 245–46.

99. G. W. Baker, *A Review of the Relative Commercial Progress of the Cities of New York and Philadelphia* (Philadelphia, 1859), pp. 22–23.

272   *Notes: The Pennsylvania Mainline*

100. George H. Burgess and Miles C. Kennedy, *Centennial History of the Pennsylvania Railroad Company, 1846–1946* (Philadelphia, 1949), p. 25.

101. Bishop, "State Works of Pennsylvania," 248–49*n*. Traffic over the trunk turnpikes from the east to Pittsburgh between 1818 and 1824 has been estimated at 30,000 tons annually, an amount that continued to grow even after the completion of the Pennsylvania mainline. Through traffic on the Erie Canal in 1837 amounted to 38,893 tons westbound, 54,219 tons eastbound; in *1845*, 42,415 tons westbound, 304,551 tons eastbound; in *1854*, 261,752 tons westbound, 1,100,526 tons eastbound. See Sadove, Transport Improvement and the Appalachian Barrier, p. 138; Reiser, *Pittsburgh's Commerical Development*, p. 77.

102. Sadove, Transport Improvement and the Appalachian Barrier, pp. 138, 154–55, 356–57; Hunter, *Steamboats on the Western Waters*, pp. 483–84; Emory R. Johnson, et al., *History of Domestic and Foreign Commerce of the United States* (2 vols.; Washington, 1915), I, 236–37; Dean, "The Theory of Geographic Location of Economic Activities," pp. 198–99, 211–13; Parkins, "The Development of Transportation in Pennsylvania," XIV, 159; Israel D. Andrews, *Report . . . on the Trade and Commerce of the British North American Colonies, and upon the Trade of the Great Lakes and Rivers*, Senate Executive Doc. 112, 32d Congr., 1st Sess. (Washington, 1854), p. 262. Sadove uses the western division of the National Road as the approximate dividing line between the northern west, which sent its exports via the Great Lakes to New York City, and the southern west, which continued to use the Mississippi route. See Table 3 of Chapter V, below, for the percentage distribution of the commodity trade of the west among the routes, 1835–1853, as computed from Sadove's figures.

103. The only total tonnage figures available include the trade of the branches of the Pennsylvania State Works and the local trade entering Philadelphia from the Susquehanna and the Delaware. For the figures, see Sadove, Transport Improvement and the Appalachian Barrier, p. 152, who took them from Michel Chevalier, *Histoire et Description des Voies de Communication aux Etats-Unis* (2 vols.; Paris, 1840–41).

104. Henry K. Strong, "Main Line of State Works of Pennsylvania," *The Merchants' Magazine*, XIII (August, 1845), 127–38.

105. Reiser, *Pittsburgh's Commercial Development*, p. 106.

106. Victor S. Clark, *History of Manufactures in the United States* (3 vols.; New York, 1949), I, 349–50.

107. Baker, *Review of the Relative Commercial Progress of . . . New York and Philadelphia*, p. 34; and the letter from S. Moylan Fox

in Philadelphia Select and Common Councils, *Reports of the Joint Special Committee . . . Relative to the Pennsylvania Railroad Company* (Philadelphia, 1846), p. 20.

108. Johnson, *History of Domestic and Foreign Commerce*, I, 238–39; Dean, "The Theory of Geographic Location of Economic Activities," pp. 211–13.

109. Johnson, *History of Domestic and Foreign Commerce*, I, 238.

110. Parkins, "Development of Transportation in Pennsylvania," XV, 8–9. For the effects of the Pennsylvania Railroad on Philadelphia's rivalry with Baltimore, see Livingood, *Philadelphia-Baltimore Trade Rivalry*, p. 151.

111. Jackman, *Development of Transportation*, II, 524–27.

112. Job R. Tyson, *Letters on the Resources and Commerce of Philadelphia* (Philadelphia, 1852), p. 14.

113. Bishop, "State Works of Pennsylvania," pp. 246–48; Andrews, *Report . . . on the Trade and Commerce of the British North American Colonies*," p. 264; *Niles' Weekly Register*, LX (May 22, 1841), 188; *Register of Pennsylvania*, VII (April 1831), pp. 218–19; Livingood, *Philadelphia-Baltimore Trade Rivalry*, p. 132.

114. Pennsylvania House *Journal*, 1838/39, III, 514, cited in Bishop, "State Works of Pennsylvania," 247n.

115. *Ibid.*, 248n.

116. Solomon W. Roberts, *Reminiscences of the First Railroad over the Allegheny Mountains* (Philadelphia, 1879), p. 76. See also, Burgess and Kennedy, *Centennial History of the Pennsylvania Railroad Company*, pp. 20–21.

117. Parkins, "Development of Transportation in Pennsylvania," XIV, 164. See also, Burgess and Kennedy, *Centennial History of the Pennsylvania Railroad Company*, p. 21.

118. Sadove, Transport Improvement and the Appalachian Barrier, p. 152, citing Frederick A. Cleveland and Fred Wilbur Powell, *Railroad Promotion and Capitalization in the United States* (New York, 1909), p. 102.

119. See the letters from Edward Miller and S. Moylan Fox, engineers, on pp. 19 and 20.

120. For such criticisms, see Andrews, *Report . . . on the Trade and Commerce of the British North American Colonies*, p. 264; and *American Railroad Journal*, XXV (January 31, 1852), p. 67.

121. Taylor, *Transportation Revolution*, pp. 80–83.

122. Johnson, *History of Domestic and Foreign Commerce*, I, 237.

123. It should be noted, however, that the railroads west of Albany were not yet physically connected to the Hudson River Railroad. See Albion, *Rise of New York Port*, pp. 384–85; also, Robert G. Albion,

"New York Port and its Disappointed Rivals, 1815–1860," *Journal of Business and Economic History*, III (August, 1931), p. 617; Parkins, "Development of Transportation in Pennsylvania," XV, 12; Walter Isard, The Economic Dynamics of Transport Technology (Unpublished PH.D. dissertation, Harvard University, 1943), p. 52. Isard speculates that, if the trans-Allegheny railroad had appeared earlier, "the economic ascendancy of New York City would have been less marked; perhaps her industrial and commercial supremacy would not have been achieved at all." However, New York had other advantages.

124. Tyson, *Letters on the Resources and Commerce of Philadelphia*, p. 14.

125. Letter to Governor John A. Schulze, November 3, 1824, in Pennsylvania *House Journal*, 1824/25, II, 330.

126. *Pennsylvania Archives*, 4th Series, V, 548–49.

127. *Internal Improvement—No. IV. Railways and Canals.*

128. *Brief View of the System of Internal Improvement in the State of Pennsylvania*, pp. 20–21.

129. See Rubin, Imitation by Canal or Innovation by Railroad, *passim*.

III. IMPROVEMENTS WITHOUT PUBLIC FUNDS: THE NEW JERSEY CANALS

This chapter is based on H. Jerome Cranmer, The New Jersey Canals; a Study of the Role of Government in Economic Development (Unpublished PH.D. dissertation, Columbia University, 1955).

1. Newark *Centinel of Freedom*, December 28, 1824.

2. An early exception to this "hands-off" policy was the Newark Turnpike Company whose 1804 charter contained provision for a legislative subscription to stock. In 1834 the Legislature authorized the exchange of this stock for shares in the New Jersey Railroad and Transportation Company. Later, in order to protect her option to purchase additional shares in the company, the state's Legislature granted a $100,000 loan to the New Jersey Railroad and Transportation Company. See also John W. Cadman, *The Corporation in New Jersey: Business and Politics, 1761–1875* (Cambridge, Mass., 1949), pp. 47, 60, 60n, 266, 291.

3. This point of view is vigorously set forth in the article from which these quotations are taken. Cornelius C. Vermeule, "Early Transportation in and about New Jersey," New Jersey Historical Society *Proceedings* IX, 2 (April, 1924), pp. 106–24.

4. See pp. 117 ff.

5. "John Pintard's College Oration," Princeton, N. J., December 1774. Cited in John Rutherford, First Thoughts on Reading the Report of Commissioners on the Morris Canal, 1823. (MS in New Jersey Historical Society Library).
6. New Brunswick *Times*, March 30, 1820.
7. Trenton *True American*, August 17, 1822.
8. New Brunswick *Fredonian*, August 1, 1822.
9. New Brunswick *Times*, August 1, 1822.
10. Trenton *True American*, June 28, 1823.
11. New Brunswick *Times*, as reprinted in Newark *New Jersey Weekly Eagle*, July 5, 1823.
12. New Brunswick *Fredonian*, November 20, 1823.
13. Trenton *Emporium*, October 25, 1823.
14. "Among the Nail Makers," *Harpers New Monthly Magazine*, XXL (July, 1860), 7.
15. Morristown *Palladium of Liberty*, June 27, 1822.
16. *Ibid.*, July 4, 11, 18, and August 29, 1822.
17. *Ibid.*, August 29, 1822.
18. *Ibid.*, September 19, 1822.
19. De Witt Clinton to New Jersey Canal Commissioners, October 27, 1823, *Niles' Weekly Register*, XXV, 220 ff.
20. *Report of the Commissioners Appointed by the Legislature of the State of New Jersey, for the Purpose of Exploring the Route of a Canal to Unite the River Delaware, near Easton, with the Passaic, near Newark* (Morristown, N. J., 1823).
21. John Rutherford, First Thoughts on Reading the Report of Commissioners on the Morris Canal (MS).
22. Morristown *Palladium of Liberty*, November 20, 1823.
23. Thomas F. Gordon, *Gazetteer of the State of New Jersey* (Trenton, 1822), pp. 23–24.
24. Newark *Centinel of Freedom*, January 3, 1825.
25. George P. M'Culloch to Cadwallader Colden as reprinted in *New Jersey Biographical Encyclopedia* (New York, 1885), p. 287.
26. "Report of the Committee to Whom Was Referred the Report of the Commissioners Appointed for the Purpose of Ascertaining the Expediency and Practicability of a Canal, from the Delaware to the Raritan Rivers, and for Other Purposes" (New Jersey 49th Legislature (1824) *Votes and Proceedings*, pp. 119–23).
27. No copy of the original plan has been found. However, extracts and summaries were printed in many New Jersey papers. See Trenton *Emporium*, December 4, 1824, and Trenton *Federalist*, December 6, 1824.
28. "Report of the Joint Committee to Examine and Enquire into

the Expediency of Constructing the Two Canals," printed in the Morristown *Palladium of Liberty,* January 6, 1827.

29. Newark *Centinel of Freedom,* December 28, 1824.

30. *Ibid.,* January 17, 1825.

31. New Jersey *Acts of the Legislative Council and General Assembly* (Trenton, 1824).

32. *Ibid.*

33. Trenton *Federalist,* December 3, 1827.

34. *Ibid.,* December 17, 24, and 31, 1827.

35. Trenton *Emporium* and Trenton *True American,* November 14 and 21, 1829.

36. "Report of the United States Army Engineers on the Morris Canal" as appended to *Report of the Commissioners Appointed by the Legislature of the State of New Jersey, for the Purpose of Exploring the Route of a Canal to Unite the River Delaware, near Easton, with the Passaic, near Newark* (Morristown, N. J., 1823).

37. Newark *Centinel of Freedom,* August 13, 1822.

IV. CYCLES OF CANAL CONSTRUCTION

1. Alexis de Tocqueville, *Democracy in America,* ed. by Phillips Bradley (New York, 1954) II, 166.

2. We made several graphical regressions of canal expenditures on miles completed and failed to obtain a close "fit," after introducing time lags of varying lengths.

The estimates of canal expenditures or the value of construction work put into place were made by H. Jerome Cranmer and presented in his paper on "Canal Investment, 1815–1860," in *Trends in the American Economy in the Nineteenth Century* (Studies in Income and Wealth, XXIV; Princeton, N. J., 1960), pp. 547–64. The figures on the number of canal miles completed annually are taken from Walter Isard, "A Neglected Cycle: the Transport Building Cycle," *Review of Economic Statistics,* XXIV (1942), 149–58. See "Estimates of Canal Investment" (pp. 208–13).

3. For a description of the most important of these adjustments see Harvey H. Segal, Canal Cycles, 1834–1861: Public Construction Experience in New York, Pennsylvania, and Ohio (Unpublished PH.D. dissertation, Columbia University, 1956) pp. 108, 119–20, 211–12, 243, 278.

4. The eleven business cycles, from 1815 through 1860, are measured from peak to peak. For the period 1815 to 1834 we determined the business cycle turning points from Williard Long Thorp, *Business*

Annals (New York, 1926). The dates for the period from 1834 through 1860 are from the business cycle chronology of the National Bureau of Economic Research, Inc. See *Historical Statistics of the United States, 1789–1945* (Washington, 1949), p. 320.

5. The prism sizes of both canals were roughly approximate: $40' \times 28' \times 4'$ (widths at the surface and bottom and the depth in feet) for Erie and $40' \times 26' \times 4'$ for the Ohio and Erie. However, the maximum elevation on the Ohio and Erie was 1,207 feet as against only 656.6 feet on the Erie. T. C. Purdy, "Report on the Canals of the United States," *Report on the Agencies of Transportation in the United States* (10th Census; Washington, 1883), pp. 22, 24.

6. Throughout this chapter a number of references to events in New York, Pennsylvania, Ohio, Illinois, and Maryland will not be documented by footnotes. Supporting evidence can be found in Harvey H. Segal, Canal Cycles, 1834–1861 (MS), and Segal, The Mixed Enterprise System of Internal Improvements in Maryland (Unpublished MS).

7. For evidence drawn from the time-expenditure pattern of a number of nineteenth century canals, see Cranmer, "Canal Investment, 1815–1860."

8. See Wayland Fuller Dunaway, *History of the James River and Kanawha Company* (New York, 1922), pp. 50–58, 86.

9. Construction work on two important anthracite canals in Pennsylvania, the Union and the Lehigh Coal and Navigation, began in 1819 and 1816 respectively, well before the success of the Erie was assumed.

10. The investment in the public canals of New York, Pennsylvania, and Ohio for the period 1815 through 1860 was $101.7 million: $28.1 million was invested during the first canal cycle, 1815–34, and the remaining $74.6 million during the second and third canal cycles, 1834–60. The figures for the period 1815–33 are taken from H. Jerome Cranmer's work sheets.

11. On the reliance of the states upon loans for internal improvement, see the contemporary analysis of Alexander Trotter, *Observations on the Financial Position and Credit of Such of the States of the North American Union as Have Contracted Public Debts* (London, 1839), pp. 72–88.

About $8 million of the net revenues earned by the Erie Canal were devoted to its enlargement between 1833 and 1856. In Pennsylvania $1.3 million of the funds invested in canal construction between 1846 and 1858 were derived from tax revenues. Elsewhere in the country the amount of public funds derived from net canal revenues or taxes was small, probably net in excess of $.7 million.

12. We estimated the amount of state canal bonds sold to foreign investors in the period 1815–60 as follows. Some of those issued by Maryland were in payment for stock subscriptions to the Chesapeake and Ohio Canal Company.

State	Amount (in millions of dollars)
New York	12
Pennsylvania	24
Maryland	7
Illinois	4
Indiana	7
Ohio	5
Virginia	3

SOURCES: New York and Pennsylvania: Segal, Canal Cycles, 1834–1861 (MS), pp. 14, 59, 131; Maryland: Segal, Canal Cycles, 1834–1861 (MS), and Segal, The Mixed Enterprise System of Internal Improvements in Maryland (MS); Illinois: John H. Krenkel, *Illinois Internal Improvements, 1818–1848* (Cedar Rapids, Iowa, 1958), pp. 114–22, 187–88; Indiana: Logan Esarey, "Internal Improvements in Early Indiana," *Indiana Historical Society Publications*, V, No. 2 (1912), 128, and Reginald C. McGrane, *Foreign Bondholders and American State Debts*, (New York, 1935), p. 140; Ohio: Estimates were made from information provided in the *Report of the Board of Commissioners Appointed by the Act of March 12, 1845, to Examine the Books, Accounts and Proceedings of the Board of Canal Fund Commissioners*, Ohio Executive Documents, 1845–46, Part 2, Table D, pp. 652–59; Virginia: Trotter, *Observations on the Financial Position of . . . the States*, p. 207, states that $2,032,400 of the improvement debt was owned by foreign bondholders in 1838. We have assumed that $3 million of Virginia's $6.8-million canal debt, incurred between 1815 and 1860, was purchased by foreign investors.

13. See Guy Stevens Callender, "The Early Transportation and Banking Enterprises of the States in Relation to the Growth of Corporations," *Quarterly Journal of Economics*, XVIII (1903), 151–52.

14. In the London money market the availability of credit was ultimately determined by the specie holdings of the Bank of England, which tended to vary its discount rate accordingly. Until 1834, the Second Bank of the United States, by virtue of its position as the depository for federal treasury funds, was able to relax or tighten credit. See Walter Buckingham Smith, *Economic Aspects of the Second Bank of the United States* (Cambridge, Mass., 1953), Chapter III. After 1834, changes in the United States money supply, banknotes plus deposits, were accomplished principally through compen-

sating variations in the reserve ratios of the state banks. These ratios varied in response to both internal and external disturbances, such as the desire of the American public to hold more gold or changes in the balances of trade and foreign capital inflows. For an analysis see George Macesich, Monetary Disturbances in the United States, 1834–1845 (Unpublished PH.D. dissertation, University of Chicago, 1958), especially Chapters II and III.

15. The canal loans constituted only a portion of a massive inflow of foreign—mainly British—capital which began in the late 1820s and extended until 1840. For annual estimates of that inflow of capital, which encompass both long- and short-term credit, see Douglass C. North, "The United States Balance of Payments 1790–1860," in *Trends in the American Economy in the Nineteenth Century,* Charts IV and V.

16. The difficulty stems largely from the fact that much of the canal construction activity took place in sparsely settled areas for which there is little quantitative information on wages and the costs of construction materials. Since the public canals were built almost solely on a contract basis, state officials were concerned only with the levels of total costs; and, on the basis of such scanty information as their records provide, it is not possible to construct a reliable index.

17. Nathan Miller, The Enterprise of a Free People: the Erie Canal and the Erie Canal Fund in the Economy of New York State, 1815–1837 (Unpublished PH.D. dissertation, Columbia University, 1959), as cited in Carter Goodrich, *Government Promotion of American Canals and Railroads, 1800–1890,* (New York, 1960) p. 54.

18. *Historical Statistics of the United States, 1789–1945* (Washington, 1949), p. 131.

19. Champlin to Williams, November 17, 1823, reprinted in John Kilbourn, *Public Documents Concerning the Ohio Canals* (Columbus, 1836?), p. 94.

On December 15, 1823, another New York correspondent, William Bayard, wrote: "I am inclined to believe that foreign capitalists would with avidity adventure a loan of this nature." *Ibid.,* p. 92. De Witt Clinton, at about the same time, assured Williams that Ohio could borrow abroad since the federal government, the "greatest borrower is . . . out of the market." *Ibid.,* p. 86.

20. Miller cited in Goodrich, *Government Promotion of American Canals and Railroads,* p. 54, and *Hazard's Register of Pennsylvania,* XII, 58.

21. By 1834 more than seventy-eight percent of all New York canal bonds were held abroad. We therefore assume three quarters of the $8.5 million New York canal debt of 1832 was foreign-owned. The

Pennsylvania canal debt in 1832, exclusive of issues used for railroad construction, was about $14 million. This was derived by deducting $1.7 million for the total debt of $15.8 million.

22. The estimates of investment are in current prices, but, if the series was deflated, 1840 would very probably be the peak year.

23. The value of cotton exports increased by 141 percent between 1826–30 and 1836–40, George Rogers Taylor, *The Transportation Revolution, 1815–1860* (New York, 1951), p. 451. Total cotton production increased from 732 million bales in 1826 to 1,976 million in 1839. *Historical Statistics of the United States*, p. 109.

24. Walter Buckingham Smith and Arthur Harrison Cole, *Fluctuations in American Business, 1790–1860* (Cambridge, Mass., 1935), pp. 185, 159.

25. Macesich, Monetary Disturbances in the United States, 1834–1845 (MS), p. 6.

26. *Historical Statistics of the United States*, p. 263.

27. North, "The United States Balance of Payments, 1790–1860," *Trends in the American Economy in the Nineteenth Century*, Table XI.

28. R. C. O. Matthews, *A Study in Trade Cycle History: Economic Fluctuations in Great Britain, 1833–1842* (Cambridge, 1954), p. 55.

29. For an account of the North American Banking and Trust Company which was established by British interests for the principal purpose of extending loans in the United States, see Joseph Dorfman, "A Note on the Interpenetration of Anglo-American Finance, 1837–1841," *Journal of Economic History*, XI, (1951), pp. 140–47.

30. The level of yields on consols was low, and there was a downward trend between 1833 and 1838 that was only briefly interrupted during the cyclical disturbances of 1836–37. Yields declined from an average of 3.42 percent in 1833 to about 3.18 percent in 1838. Matthews, *A Study in Trade Cycle History*, p. 188. During this period the American state bonds yields in London ranged from 4.75 to 6.50 percent.

31. They include the state works of New York, Pennsylvania, Ohio, Indiana, and Illinois, plus the Chesapeake and Ohio and the James River and Kanawha canals, which were largely financed by the states of Maryland and Virginia.

32. The new projects included the 112-mile Erie extension, which was to link Pittsburgh with Lake Erie; the North Branch extension, a 90-mile line that was to provide a link with the New York state canal system; and the West Branch extension, which, if completed, would have connected the Susquehanna and Allegheny Rivers.

33. The projects were as follows: Miami extension, 128 miles; Wabash and Erie extension, 88; Hocking, 56; Walhonding and Mohican, 25; Muskingum, 91; Warren, 20; total, 408 miles.

34. Based on Esarey, "Internal Improvements in Early Indiana."

35. H. Jerome Cranmer estimates construction outlays for the period 1833 through 1835 at $.6 million.

36. Ohio had agreed to construct the 88-mile section of the Wabash and Erie lying between the state line and Toledo in return for a portion—292,224 acres—of the federal lands that had been granted to Indiana.

37. Based on Krenkel, *Internal Improvements in Illinois*.

38. Based upon Dunaway, *History of the James River and Kanawha Company*.

39. Ralph W. Hidy, *The House of Baring in American Trade and Finance: English Merchant Bankers at Work, 1763–1861* (Cambridge, Mass., 1949), pp. 262, 266.

40. In 1837 the Bank of the United States acquired a one-fourth interest in the Morris Canal and Banking Company, and Nicholas Biddle's cousin, Edward R. Biddle, became the president of the New Jersey corporation. Smith, *Economic Aspects of the Second Bank of the United States*, p. 211. The canal bonds purchased by the Bank of the United States and the Morris Canal and Banking Company were as follows: Pennsylvania, *1836–41*, $5.8 million; Illinois, *1837* and *1839*, $2.2 million; Indiana, *1836–37*, $7.7 million; Maryland, *1839*, $1.0 million; total, $16.7 million. (Pennsylvania: Segal, Canal Cycles, 1834–1861 (MS), pp. 171–72; Illinois: Smith, *Economic Aspects of the Second Bank of the United States*, p. 119, and Krenkel, *Internal Improvements in Illinois*, p. 115; Indiana: Esarey, "Internal Improvements in Early Indiana," p. 128; Maryland: "Governor's Annual Message," *Maryland Public Documents*, 1839, p. 7.)

41. For an account of them, see Smith, *Economic Aspects of the Second Bank of the United States*, pp. 195–202, and *Report of the Committee of Investigation Appointed at the Meeting of the Stockholders of the Bank of the United States, Held Januray 4, 1841* (Philadelphia, 1841), pp. 19–25.

42. By November, 1839, more than $5 million of the state bonds had been pledged for loans at fractions of their par value. Smith, *Economic Aspects of the Second Bank of the United States*, p. 218.

43. See Sir John Clapham, *The Bank of England: a History* (New York, 1945), II, pp. 151–53, 429.

44. *Ibid.*, p. 147. At the time sixty-day bills on London were selling at 19.75 percent premium in New York. Smith and Cole, *Fluctuations in American Business*, p. 190.

45. Construction outlays on the Chesapeake and Ohio Canal rose from $572,000 in 1836 to $857,000 in 1837.

46. The passage that follows is based upon Leland Hamilton Jenks, *The Migration of British Capital to 1875* (New York, 1927), pp. 95–98.

47. Arthur D. Gayer, Walt W. Rostow, and Anna Jacobson Schwartz, *The Growth and Fluctuations of the British Economy, 1790–1850: an Historical, Statistical and Theoretical Study of Britain's Economic Development* (Oxford, 1953), I, 296–97 and Chart 80-b, p. 300. According to Matthew's *A Study of Trade Cycle History*, p. 201, the average market rate of discount on bills rose from 3.2 to 6.0 percent between the first and third quarters of 1839.

48. Quoted from the "Report of the Fund Commissioners of Illinois, December 13, 1839," in Smith, *Economic Aspects of the Second Bank of the United States*, p. 199; also, Krenkel, *Internal Improvements in Illinois*, p. 115.

49. Peabody to Washington, May 16, 1839, quoted in "Minority Report of the Committee on Internal Improvements . . . March 1, 1841," *Maryland Public Document*, 1841, Document 17, pp. 28–29.

50. Esarey, "Internal Improvements in Early Indiana," pp. 108–9. A classification act was passed by the Legislature, authorizing the public works commissioners to establish priorities in the allocation of the available funds.

51. The proportion of New York canal bonds held abroad dropped from 76.5 percent in 1834 to 59.1 percent in 1843.

52. They amounted to nearly $4 million in New York, Pennsylvania, and Ohio.

53. "Governor's Annual Message," *New York Assembly Documents*, 1840, No. 1, p. 24.

54. Esarey, "Internal Improvements in Early Indiana," pp. 143–46.

55. Esarey "Internal Improvements in Early Indiana," pp. 109, 151; Krenkel, *Internal Improvements in Illinois*, p. 121.

56. This possibility is insinuated in William Gouge's contemporary account of the crisis, which began in January, 1842, with the failure of the Girard Bank and then engulfed the Bank of Pennsylvania. In commenting upon the run on the Girard Bank, Gouge concludes: "Thus in a few hours did the 'best currency in the world' cease to be a currency at all, not through a panic seizing the people, but simply through a panic seizing the officers of the Northern Liberties Bank." William Gouge, *The Journal of Banking from July 1841 to July 1842* (Philadelphia, 1842), p. 247. Note dated February 2, 1842.

57. The total of canal currencies in circulation of the spring of 1842 was as follows: State of Indiana, $1,200,000; State of Ohio,

$701,000; State of Pennsylvania, $2,187,000; the Chesapeake and Ohio Canal Company, $712,000; the James River and Kanawha Company, $205,000; the Susquehanna and Tidewater Company, $410,000; total, $5,415,000. (Sources: The Chesapeake and Ohio and Susquehanna and Tidewater: Company reports cited in Segal, Canal Cycles, 1834–1861 (MS), and Segal, The Mixed Enterprise System of Internal Improvements in Maryland (MS); Ohio and Pennsylvania: Segal, Canal Cycles, 1834–1861 (MS); Indiana: Esarey, "Internal Improvements in Early Indiana"; James River and Kanawha: James River and Kanawha Company, *Annual Report of the President to the Stockholders,* 1842.)

58. According to Macesich, Monetary Disturbances in the United States, p. 6, the money supply—bank notes plus deposits—declined by eighteen percent between 1839 and 1842.

59. They included the state canals of New York, Pennsylvania, Indiana, and Illinois, together with the Chesapeake and Ohio and the James River.

60. Reprinted in the *Baltimore Sun,* February 4, 1840, p. 1.

61. The sonnet was probably written in 1845. See William Wordsworth, *Poetical Works,* ed. by Edward Dowden (London, 1893), IV, pp. 279, 385–86.

62. For accounts of those efforts, see McGrane, *Foreign Bondholders and American State Debts,* Chapters IV-VII, and Hidy, *The House of Baring,* Chapter 11.

63. Esarey, "Internal Improvements in Early Indiana," p. 109.

64. The foreign loans were as follows: James River and Kanawha, $900,000; Illinois and Michigan, $1.3 million; and Wabash and Erie, $900,000; total, $3.1 million. (Sources: James River and Kanawha: Hidy, *The House of Baring,* pp. 424–25; Illinois and Michigan: *Report of the Board of Trustees of the Illinois and Michigan Canal,* 1847–49, *Illinois Reports,* 1846, 1849; Wabash and Erie: Reginald C. McGrane, *Foreign Bondholders and American State Debts,* pp. 140–41.)

Both the Illinois and Michigan and the Wabash and Erie canals were deeded to trusteeships of British bondholders who provided the bulk of the funds for their completion in the later 1840s and early 1850s.

65. "Governor's Message," *Maryland Public Documents,* 1852, Document A, p. 17.

66. The Illinois and Michigan was completed at a cost of $1.3 million, and the Wabash and Erie at $3.6 million. Data for the latter were found in Cranmer's work sheets.

67. See Carter Goodrich, "The Virginia System of Mixed Enterprise: a Study of the Planning of Internal Improvements," *Political Science*

Quarterly, LXIV (1949), 383–84; and James River and Kanawha Company, *Annual Report of the President to the Stockholders*, 1856 and 1857.

68. Richard A. Easterlin, "Interregional Differences in Per Capita Income, Total Income and Population in the United States, 1840–1950," in *Trends in the American Economy in the Nineteenth Century*, pp. 97–8.

69. Robert E. Gallman. "Commodity Output in the United States, 1839–1899," in *Trends in the American Economy in the Nineteenth Century*.

V. CANALS AND ECONOMIC DEVELOPMENT

1. Michel Chavalier, *Society, Manners and Politics in the United States, Being a Series of Letters on North America* (Boston, 1839) Letter XXI, Buffalo, July 9, 1835, pp. 274–75. His *Histoire et description de voies des communication aux Etats Unis et des travaux qui en dependent* (Paris, 1840–41) is still the best general account of American transport facilities in the middle and later 1830s.

2. *The Wealth of Nations* (Modern Library Edition), p. 382.

3. Albert Gallatin, *Report of the Secretary of the Treasury on the Subject of Public Roads and Canals* (Washington, D.C., 1808).

4. New York State, *Messages from Governors* (Albany, 1909), II, p. 1006.

5. *Ibid.*, p. 1007.

6. Annual Message, January, 1826, in *Papers of the Governors, 1691–1902* (Pennsylvania Archives, Fourth Series; Harrisburg, Pa., 1900–2), V, 657.

7. Daniel Raymond, *Thoughts on Political Economy* (Baltimore, 1820), p. 290. This passage and the quotation that follows were eliminated from subsequent editions.

8. *Ibid.*, p. 294.

9. Vethake to Carey, January 26, 1825. The full letter appears in Joseph Dorfman and R. G. Tugwell, *Early American Policy: Six Columbia Contributors* (New York, 1960), pp. 160–61.

10. According to Samuel Blodget's estimates for 1805, the value of real and personal property in the United States was $2.51 billion. Of this total, land accounted for $1.66 billions or more than sixty-six percent. The ratio of land to total property values was also very high in 1850. See *Historical Statistics of the United States, 1789–1945* (Washington, D.C., 1949), pp. 1–2, Tables 1 and 2.

11. We could easily have enlarged our sample of contemporary

statements, but the number of relevant concepts bearing upon the process of economic development would not have been increased.

12. *The Wealth of Nations* (Modern Library Edition), pp. 17–21.

13. Estimates of per capita by states are not available before 1840. In that year however, the per capita incomes of the trans-Appalachian states were considerably lower than those in the New England, the Middle States, or the nation on the whole. In 1840 average per capita income in the East North Central States—Ohio, Indiana, Illinois, Michigan and Wisconsin—was $46.1, in New England $83, in the Middle Atlantic States $77, and in the country as a whole $65. There are reasonable grounds for the belief that the regional differentials before 1840 were larger. See Richard A. Easterlin, "Interregional Differences in Per Capita, Population and Total Income, United States, 1840–1950," in *Trends in the American Economy in the Nineteenth Century* (Studies in Income and Wealth XXIV; Princeton, N.J., 1961), p. 97.

14. For an excellent analysis of these interrelated developments, see Douglas C. North, "International Capital Flows and the Development of the American West," *Journal of Economic History*, XV (1956), pp. 493–505.

15. See James S. Duesenberry, "Some Aspects of the Theory of Economic Development," *Explorations in Entrepreneurical History*, III, (1950) p. 98. Duesenberry points out that in process of western settlement the incomes of eastern farmers decline.

16. For a discussion of the lag of western exports behind a rising volume of imports, see Abraham H. Sadove, Transport Improvement and the Appalachian Mountain Barrier: a Case Study in Economic Innovation (Unpublished PH.D. dissertation, Harvard University, 1950), pp. 178–79. A. L. Kohlmeier in *The Old Northwest as the Keystone of the Arch of Federal Union. A Study in Commerce and Politics* (Bloomington, Ind., 1938), pp. 95–96, presents a statistical evidence, based on census data for 1840, 1850, and 1860, that supports the hypothesis of a decline in the exportable surpluses of wheat and pork, in Ohio, Indiana, and Illinois.

17. The Mainline of the Pennsylvania public works, as we shall demonstrate, was not successful as a through route, and neither the Chesapeake and Ohio nor the James River and Kanawha reached their Ohio River destinations.

18. J. H. Clapham, *An Economic History of Modern Britain: the Early Railway Age, 1820–1850* (Cambridge, 1959), p. 80.

19. Noble E. Whitford, *History of the Canal System of the State of New York*, together with *Brief Histories of the Canals of the United*

States and Canada (Albany, 1906), Vol. I, Plate 1, Appendix to Chapter XXV.

As a canal engineer writing just before the commencement of the $104 million Erie Barge Canal project, Whitford's bias was pronounced. His enthusiasm for canals was also shared by a number of other writers before the First World War, who saw in a canal revival the possibility of more effectively regulating the monopolistic power of the railroads.

20. *Ibid.*, p. 830.
21. *Ibid.*, p. 821.
22. Thomas Senior Berry, *Western Prices before 1861: a Study of the Cincinnati Market* (Cambridge, Mass., 1943), Tables 8, 9, and 10, pp. 106, 113, and 114.
23. See note 16, above.
24. George Rogers Taylor, *The Transportation Revolution, 1815–1860* (New York, 1951), p. 133.
25. *Ibid.*, p. 137. In 1854 the average rates per ton-mile on the major canals were as follows: Erie, 1.1 cents; Pennsylvania Mainline, 2.4; Ohio and Erie, 1.0; Wabash and Erie, 1.9; Illinois and Michigan, 1.4. *Hunts' Merchants' Magazine*, XXXI (1854), 123. The Pennsylvania Mainline rate is higher because it includes railroad hauls and the cost of transferring freight.
26. Evidence supporting this assertion appears in Robert W. Fogel, Essays on the Influence of the Railroads on American Economic Growth, now in preparation.
27. For an account of the difficulties that the Middlesex Canal faced in competing against the wagon-haulers, see Christopher Roberts, *The Middlesex Canal, 1793–1860* (Cambridge, Mass., 1938), pp. 148–54, 166–70.
28. Taylor, *The Transportation Revolution*, p. 134.
29. See Avard Longley Bishop, "The State Works of Pennsylvania," Connecticut Academy of Arts and Sciences *Transactions*, XIII (1908), 248. Bishop's data are for 1836, 1844, 1845, and 1854. According to the *American Railroad Journal*, XXV (1852), 67, Philadelphia merchants found it advantageous to ship goods westward over the Erie Canal.
30. For the data on "tons from western states coming to tidewater," that is, Albany, see "Report of the Auditor of the Canal Department of the Tolls, Trade and Tonnage of the Canals of the State of New York, February 1861," *New York Assembly Documents*, 1861, No. 91, pp. 190–91.

Prior to the completion of the Ohio state canals in the early 1830s, the bulk of the eastbound Erie traffic originated in western New York. See Louis B. Schmidt, "Internal Commerce and the Development of a

National Economy before 1860," *Journal of Political Economy*, XLVII (1939), 811.

31. Robert E. Gallman, "Commodity Output in the United States, 1839–1899," in *Trends in the American Economy in the Nineteenth Century*, pp. 13–67.

32. Canal tolls on freight carried by New York railroads in the vicinity of the Erie Canal were abolished in December, 1851. Among the other factors which adversely affected the canal's tonnage were: (1) the pricing practices of the private barge operators, who frequently raised their rates when the state lowered tolls; (2) the longer term bank loans that merchants who shipped by canal were compelled to make; and (3) the formation of the New York Central Railroad, through consolidation, in 1854. See Harvey H. Segal, Canal Cycles, 1834–1861: Public Construction Experience in New York, Pennsylvania, and Ohio (Unpublished PH.D. dissertation, Columbia University, 1956), pp. 91–93.

33. Computed from *Historical Statistics of the United States: Colonial Times to 1957* (Washington, D.C., 1960), p. 12.

34. See Abraham H. Sadove, Transport Improvement and the Appalachian Barrier (MS), pp. 166–68, 187.

35. See Douglass C. North, "International Capital Flows and the Development of the American West," *Journal of Economic History*, XV, (1956), 501.

36. Arthur H. Cole, "Cyclical and Sectional Variations in the Sates of Public Lands," *The Review of Economics and Statistics*, IX (1927), 41–56, especially 49; Elbert Jay Benton, *The Wabash Trade Route in the Development of the Old Northwest* (Johns Hopkins Studies in Historical and Political Science, XXI; Baltimore 1903), pp. 92–98.

37. "Annual Report of the Board of Public Works, January 2, 1842," *Ohio General Assembly Documents*, No. 36, p. 41.

38. Israel D. Andrews, "Report on the Trade and Commerce of the British North American Colonies," *United States Senate Document* No. 112, 32d Congress, 1st Session (Washington, D.C., 1853), pp. 220–22; James William Putnam, *The Illinois and Michigan Canal: a Study in Economic History* (Chicago, 1918), pp. 102–4, 107–9; William F. Switzler, "Report on the Internal Commerce of the United States, Part II of Commerce and Navigation, Special Report on the Commerce of the Mississippi, Ohio and Other Rivers, and of the Bridges Which Cross Them," *Treasury Department Document* No. 1,039b (Washington, D.C., 1888), p. 211.

39. *Annual Report of the Auditor of the Canal Department . . . on the Canals of New York, February, 1861* (Albany, 1861), pp. 224–25.

40. A. H. Sadove, Transport Improvement and the Appalachian Barrier (MS), pp. 197–200. Sadove's tonnage series on western imports and exports were developed from material presented in A. L. Kohlmeier, *The Old Northwest*, and various federal government documents. His exports totals over the northeastern route for the benchmark years 1835 through 1853 are considerably lower than the tonnage reaching tidewater over the Erie Canal. It is possible, therefore, that Table 4 understates the change in the shares of the northeastern route.

41. The increase in trade over the three routes, 1835–53, as measured in thousands of short tons shipped in both directions, is as follows (*ibid.*, p. 197):

Year	Northeastern	Southern	Eastern	Total
1835	64	168	38	270
1853	1,168	543	166	1,877

42. See Berry, *Western Prices before 1861*, pp. 122–29, 564.

43. *Ibid.*, pp. 122–23.

44. Total manufacturing employment in the two states increased from 120,200 in 1820 to 279,100 in 1840. *United States Census*, 1820 and 1840.

45. The following paragraphs rely heavily upon Chester Lloyd Jones, *The Economic History of the Anthracite-Tidewater Canals* (University of Pennsylvania Series in Political Economy and Public Law, No. 22; Philadelphia, 1908).

46. We have omitted the Morris Canal, because it never developed a really substantial coal trade, despite the anticipations of its builders. Its coal tonnage, which was drawn from the Lehigh Navigation, did not exceed 291,000 tons in the ante-bellum period. *Ibid.*, p. 122.

47. *Historical Statistics of the United States, 1789–1945*, p. 142.

48. The intense rivalry between the Schuylkill Navigation Company and the Reading Railroad began in 1844. Jones, *Economic History of the Anthracite-Tidewater Canals*, pp. 136–38.

49. *Ibid.*, pp. 29, 35, 36, 86, 155–56. In computing the totals, we excluded the tonnage on the Delaware Division Canal that came largely from the Lehigh Navigation, in order to avoid double counting.

50. The shares of the canals are based on a comparison of canal tonnage with data on shipments from the Schuylkill, Lehigh and Wyoming County coal fields of Pennsylvania. See U. S. Geological Survey *Mineral Resources of the United States* (Washington, D.C., 1883), p. 13.

51. Computed from data for 1820 presented in Whitford, *History of the Canal System of the State of New York*, I, pp. 920–21, and the "Annual Report of the Comptroller, 1847, Statement No. 8," *New*

York Assembly Documents, 1847. The wholesale price deflators for 1820 and 1847 are those prepared by Professor George Rogers Taylor, see "Historical and Comparative Rates of Production, Productivity, and Prices," in *Employment, Growth and Price Levels,* Part 2, Hearings before the Joint Economic Committee of the Congress of the United States, April 7, 8, 9 and 10, 1959 (Washington, D.C., 1959), p. 395.

52. See Laura Randall Rosenbaum, The Effect of the Erie Canal on the Economic Growth of New York State, 1820–1850: a Study in Location Theory (Unpublished master's thesis, University of Massachusetts, 1959), p. 54.

53. The increases in population, 1820–40, were as follows:

Area	Percentage Increase
Erie Canal Counties	99.1
New York County	152.8
Kings County	325.8
Others: excluding Kings and New York	56.8
State	76.9

SOURCE: *United States Census,* 1820 and 1840.

54. Moving northward from Cincinnati the percentages of the population engaged in manufacturing by counties are as follows: Hamilton, 13.9; Butler, 5.8; Warren, 1.7; Montgomery, 3.6; Miami, 7.1; Shelby, 1.0. Computed from *United States Census,* 1840.

55. New York Senate Journal, 1827, p. 173.

56. Dorfman and Tugwell, *Early American Policy,* p. 340.

57. "Eleventh Annual Report of the Board of Public Works, February, 1847," *Ohio Executive Documents,* 1847, p. 104.

58. For New York evidence relating to this point see, "Report of the Auditor of the Canal Department . . . on the Canals of New York, February, 1861," pp. 22–23, 27, where it appears that for flour the carriers' charges varied much more frequently and often in the opposite direction from the changes in toll rates.

59. Data on gross revenues and operating expenditures for the Lehigh, Schuylkill, Delaware and Hudson, and Delaware Division lines are found in Jones, *The Economic History of the Anthracite Tidewater Canals;* for the Erie, Champlain, and Oswego from Annual Report of the Auditor of the Canal Department . . . on the Canals of New York, February, 1861; for the Delaware and Raritan, H. Jerome Cranmer, The New Jersey Canals: State Policy and Private Enterprise, 1820–1832 (Unpublished PH.D. dissertation, Columbia University, 1955), p. 331.

60. On the Pennsylvania canals see, Segal, Canal Cycles, 1834–1861

(MS), pp. 214–15, and Bishop, "The State Works of Pennsylvania," pp. 278–85; on the Ohio canals, C. P. McClelland and C. C. Huntington, *History of the Ohio Canals, Their Construction, Cost, Use and Partial Abandonment* (Columbus, Ohio, 1905), pp. 168–71, and *Ohio Senate Journal*, 1848–49, pp. 73–75; on the Wabash and Erie, Benton, *The Wabash Trade Rouie*, pp. 76–78; on the Illinois and Michigan, Putnam, *The Illinois and Michigan Canal*, p. 161; on the James River and Kanawha, Wayland Fuller Dunaway, *History of the James River and Kanawha Company, 1836–1860;* on the Chesapeake and Ohio, Walter S. Sanderlin, *The Great National Project: a History of the Chesapeake and Ohio Canal* (Baltimore, 1946), p. 309.

61. For a comprehensive treatment of the underlying theory and techniques, see Otto Eckstein, *Water Resource Development: the Economies of Project Evaluation* (Cambridge, Mass., 1958), especially Chapters II, III, and VI.

62. Following, *ibid.*, p. 56, we shall employ the following formula:

$$\frac{B}{C} = \frac{B}{O + a_{it} K}$$

where B = benefit; O = operating expenses; i = the real rate of return on capital, which we estimate at 8 percent for the ante-bellum period; t = the amortization period, assumed to be 50 years; and K = permanent investment. The term $a_{it} = .0817$ was obtained from a table, "Annuity Whose Present Value is 1," in Charles D. Hodgman, comp., *Mathematical Tables* (8th ed.; Cleveland, 1946), p. 324.

In calculating the cost of canals, O was reduced by the amount of tolls and other revenues. Where canal revenues exceeded operating costs, interest and amortization charges, the surplus was deducted from K.

63. See notes 24 and 25, above.

64. The costs were computed as follows: K, the average fixed investment for the period 1837–46, was estimated at $128.9 million from Table 2 (pp. 208–9).

O, operating expenditures are roughly estimated at $2 million on the basis of data relating to the canals of New York, Pennsylvania, and Ohio in Segal, Canal Cycles, 1834–1861 (MS), p. 298.

$a_{it} = .0817$ is explained in footnote 62 above.

Then $C = O + a_{it} K = \$2.0 + (.0817 \times \$128.9) = \$2.0 + 1.05 = \$3.05$ million.

65. See Table 13 (p. 246).

66. Information on tonnage, tolls, and operating costs relating to the Chesapeake and Ohio Canal were obtained from Sanderlin, *The Great National Project*, pp. 307, 309 and 311. The per ton-mile rate on coal charged by the Baltimore and Ohio Railroad was calculated from information in Balthasar Henry Meyer, ed., *History of Transportation in the United States before 1860* (Washington, D.C., 1917), p. 578.

The average annual cost of the canal, see note 62, above, was calculated in millions of dollars, as follows:
$C = O + a_{it} K = \$.175 + (.0817 \times \$11) = \$.175 + \$.985 = 1.05$ million.

We deduct $154 for canals revenues to arrive at an annual uncovered cost of $846,000. Benefits conferred by 27 million ton-miles of coal traffic amount to $270,000. Therefore, the remaining 12 million ton-miles of traffic would have had to confer benefits of $546,000—or an average of 4.8 cents per ton-miles.

67. Benton, *The Wabash Trade Route,* pp. 78–81, 188, 102–3.

68. Cited in Taylor, *The Transportation Revolution,* p. 174. For full citation of Andrews "Report," see note 38.

69. *Ibid.*

CONCLUSION

1. The quotations are from De Tocqueville and Gouverneur Morris (see above, pp. 169 and 27).

2. See above, p. 169.

3. De Witt Clinton to New Jersey Canal Commissioners, October 27, 1823, *Niles' Weekly Register,* XXV, 221.

4. Francis Wayland, *The Education Demanded by the People of the United States* (Boston, 1855), quoted in Joseph Dorfman, "The Principles of Freedom and Government Intervention in American Economic Expansion," *Journal of Economic History,* XIX (1959), p. 581.

5. Julius Rubin is continuing research on this point.

6. See quotation by De Witt Clinton, p. 65.

Index

Agriculture: in New Jersey, 126–27, 159; commodities, 190; surplus, 218, 221–25; commodity prices, 230
Albany, N.Y., 17–18, 20–21, 32
Albemarle and Chesapeake Canal, mileage and investment (*table*), 212
Albion, Robert G., 112, 273 *n*123
Alexandria and Georgetown Canal, mileage and investment (*table*), 212
Alexandria Canal Company, government aid to (*table*), 214
Allegheny Portage Railroad, 228
Allegheny River, 7
Appalachian Mountains: as transport barrier, 4, 7–8, 15, 68, 113, 177, 217, 221–22, 231; gap through, 16; railroad over, 112
Armroyd, George, 7, 257 *n*18
Augusta Canal, mileage and investment (*table*), 212
Aviation, 9

Bald Eagle and Spring Creek Canal, mileage and investment (*table*), 211
Baltimore, 6, 67, 68, 102
Baltimore and Ohio Railroad, 8, 105, 245
Bank of England, 179–80, 196, 278 *n*14
Bank of Pennsylvania, 187–89, 200–1, 282 *n*56
Bank of the Manhattan Company, 195
Bank of the United States, 60, 195, 197, 200, 281 *n*40
Bank of the United States, Second, 146, 179–80, 188, 190–91, 193, 278 *n*14
Banking privileges, in aid of New Jersey canals, 140–46, 162–63, 251, 254

Banks: and canal financing, 5, 179–81, 187–89, 224; and canal cycles, 191–92, 195, 205; and panic of *1839*, 197
Banks, local, 224
Baring Brothers, 187, 195–98 *passim*, 201–2
Bayard, William, 53, 279 *n*19
Benefit-cost analysis, applied to canals, 217, 241–44, 290 *n*62
Benton, Elbert Jay, 246
Bergen County, N.J., 130, 141
Berry, Thomas S., 227, 232, 286 *n*2, 288 *n*42–43
Biddle, Edward R., 146, 281 *n*40
Biddle, Nicholas: and Bank of the United States, 146, 188, 191, 281 *n*40; and second canal cycle, 195, 197
Birmingham and London Railroad, 76
Bishop, Avard L., 82, 99, 268 *n*37, 272 *n*101
Black River Canal, 192, 203, 205; mileage and investment (*tables*), 211, 246
Blackstone Canal, 178; mileage and investment (*table*), 211
Bonds, state: for canals, 150, 179–80, 182, 187–88, 191, 195, 197, 198, 278 *n*12; in second canal cycle, 194–95, 207, 281 *n*40–42, 282 *n*51
Bonus, paid for canal charter, 143–44
Boom, *see* Business cycles
Boston, 7
Bridgewater, Duke of, 2
Brindley, Joseph, 3
Bristol, England, 4
Brokerage houses, 179, 195
Brunswick Canal, mileage and investment (*table*), 212
Buffalo, N.Y., 31, 232
Burlington County, N.J., 138

Business cycles: and canal cycles, 11, 173, 190, 205–8, 250, 276 n4; and western settlement, 223–24

Busti, Paul, 56

Calhoun, John C., 130

Camden, N.J., and New Jersey canals, 136

Camden and Amboy railroad project, 152, 153, 154, 156

Canada, 26, 54, 64, 264 n84

Canal bonds, see Bonds

Canal cycles, 171–72, (table), 172; charted, 173; long, 173–83; models, 174, 181–83; and stagger period, 175–76; and sectional rivalry, 176–79; role of finance in, 179–81; monetary and fiscal impacts, 205; and employment, 205, 206; and business cycles, 205–8; investment by (table), 210; and business cycles, 250

——First, 171, 172 (chart), 176, 182–89; in Ohio, 178; percent of total investment, 186; and publicly financed projects, 186; financing construction in, 186; economic effects, 207–8; investment during, 210, 215

——Second: 172–73, 180, 182–83, 189–201; rivalry and, 179; peak year, 189, 232, 280 n22; investment during, 189–92, 210, 215, 280 n22; and banks, 191–92, 195, 205, 280 n29; and state financing, 192–201; and panic of 1839, 198–201, 250; economic effects, 206–8

——Third, 172, 173, 201–5; "echo phenomenon," 183, 250; economic effects, 207–8; investment during, 210, 215

Canal investment, see Investment

Canal "mania," see Canal movement

Canal movement, 4–5, 11, 91, 169, 176–77, 252–53

Canal "scrip," 189, 196, 198, 200, 282 n57

Canals: and decision-making, 1–10 (see also specific canals, e.g., Erie Canal); and railroads, 10–11, 77–114 passim, 244–46, 270 n64; economic importance, 11–12, 15, 216, 220–38, 284 n10, 285 n19, 286

n25; European, 19; technology, 23, 52, 127–28, 170, 174–76 (see also Engineering); exploitative or developmental, 157–66, 238, 243, 251–52; principal ante-bellum (map), 184–85; construction cost indices for, 186; investment (table), 211–13; and western settlement, 223, 230; appraisal of, 238–46; investment criteria for, 239–46, 277 n10; net revenues, 240; financial failures among, 240–41, 244–46, 246 (table), 250; ton-mileage on (table), 242; underutilization, 244; in 1850s, 244–46, 247; transport services of, 247–48; see also Construction; Great Britain; specific canals, e.g., Erie Canal

Canandaigua, N.Y., 24, 28, 29

Cape Cod Canal, 8

Capital: sources for canals, 1, 5, 34–36, 118–19, 179–80; and settlement, 223; foreign (see Foreign capital)

Capital formation, and canal expenditures, 207–8

Capital market, 11, 250

Carey, Matthew, 220, 252; founder of Philadelphia Society, 70–71; in canal-railroad debate, 78–99 passim, 113–14, 268 n39

Carondelet Canal, 6

Cash crops, 222

Catawba Canal, mileage and investment (table), 212

Cayuga and Seneca Canal, mileage and investment (table), 211

Cayuga Lake, 22, 32, 41, 42

Centinel of Freedom (Newark, N.J.), 141, 144

Champlain Canal, 45, 58, 59, 178, 260 n43; mileage and investment (table), 211; ton-mileage on, 242

Champlin, John T., 187

Charleston, S. C., 7; see also Santee Canal

Chemung Canal, 175, mileage and investment (table), 211

Chenango Canal, 244; mileage and investment (tables), 211, 246

Chesapeake and Delaware Canal, 97, 242; mileage and investment (tables), 212, 214

Chesapeake and Ohio Canal, 6–8; and

canal cycles, 176, 177–78, 196, 203; financing, 187, 188–89; mileage and investment (*tables*), 212, 214, 246; and interregional trade, 228; and railroad competition, 244–46, 286 n17
Chevalier, Michel, 216, 242, 284 n1
Chicago, 8, 112, 232
Cincinnati, 67, 178, 227, 236–37
Cincinnati and Whitewater Canal, mileage and investment (*tables*), 212, 214
Civil War, 9, 205
Clapham, J. H., 226, 285 n18
Clarke, James, 72
Cleveland, 174, 232
Clinton, De Witt: and Erie Canal project, 24, 39, 40, 52, 252; "New York Memorial" of, 54–65 *passim*, 223, 261 n60, 262 n72, 263 n65, 263 n75–76, 264 n84, 279 n19; and Pennsylvania Mainline, 96; and New Jersey canals, 130–31, 137; quoted, 218–19
Clinton, George, 33, message to the legislature, 19
Clyde Canal, 29
Coal, 135–36, 146, 159
Coal canals, 177–78, 234, 277 n9
Colden, Cadwallader, 1724 memorial by, 17–19, 20, 34, 37, 54
Columbia Canal, government aid to (*table*), 214
Commodity production, transport costs and, 217–18
Commodity "surplus," 217, 221
Commodity trade, impact of canals on, 227–32; growth of output (*table*), 229; distribution by routes (*table*), 231
Commonwealth (Pittsburgh), 28
Company's Canal, mileage and investment (*table*), 213
Comparative capability, 65–66, 114
Congress, and Erie Canal project, 39–47, 60
Connecticut, 178; canal mileage and investment (*table*), 211
Consols, 192, 280 n30
Constitution, bank amendment advocated, 46
Constitutional Convention of 1846 (New York), 204

Construction: measuring, 170–71, 276 n2–3; cost indices for, 186; volume, 195; and third canal cycle, 203–5; and economic activity, 206–8
Consumption, relation to transport costs and investment, 219
Cotton, 182, 195, 280 n23
Cranmer, H. Jerome, 115; cited, 170–71, 172, 208, 210, 233, 242, 272, 274, 276 n2, 277 n10
Credit, state, for improvement projects, 161
Crooked Lake Canal, mileage and investment (*table*), 211
Cumberland and Oxford Canal, mileage and investment (*table*), 211
Cumberland Road, 57
Customs revenues, 36
Cycles, see Business cycles; Canal cycles

Dayton, Ohio, 178
Debts, on public canals, 171, 195, 278 n12
Decision-making, see specific canals, e.g., Erie Canal
Delaware, canal mileage and investment (*table*), 212
Delaware and Hudson Canal, 242; mileage and investment (*tables*), 211, 214
Delaware and Ohio, proposed canal, 67
Delaware and Raritan Canal: decision-making and, 8, 10, 116–56, 265 n7; summary of canal operation, 156; merger with Camden and Amboy company, 156; exploitative, 159; significance of, 160, 163; mileage and investment (*table*), 211; ton-mileage on, 242
Delaware Division Canal, 149, 175, 204; mileage and investment (*table*), 211
Delaware River, 116, 118
Democratic Party, 52, 58, 62, 102, 192
Depression of *1842*, 105
Development, economic, see Economic development
Developmental canals, see Canals
De Witt, Simeon, 27, 31, 40
Domestic markets, development of, 218

Drehr's Canal, mileage and investment (table), 212
Duane, William J., 82
Dunlap, Thomas, 197
Dupin, Baron, 2, 4

Easterlin, Richard A., 206
Eastern Division Canal (Pennsylvania Mainline), mileage and investment (table), 212
"Echo phenomenon," 183, 201–5
Economic development, and canals, 1, 11–12, 15, 205, 216–49, 251
Eddy, Thomas, 39, 40, 50, 53, 263 n76
Elizabeth-town, N.J., 130
Ellicott, Joseph, 31, 40, 50, 58, 262 n66, 262 n72
Embargo Act, 44
Employment, and canals, 205, 206, 224–25, 234–38; occupational distribution of (tables), 236, 237
Emporium (Trenton, N.J.), 124, 275 n27
Engineering, 2–5; and Erie Canal, 22, 49, 65, 174, 257 n14, 263 n72; and Morris Canal, 130
England, 50, 58; railroads in, 75–77, 79–81, 84, 85, 87, 94, 95; "canal mania" in, 125; *see also* Great Britain
Entrepreneurs, 224
Erie Barge Canal, 9, 285 n19
Erie Canal, 10, 158, 174; influence, 6–8, 68, 75, 91, 94, 112, 176–77, 190, 224, 226, 228–32, 235–38; decision-making and, 15–66, 258 n3, 258 n8, 263 n79; financial success, 11, 240; engineering innovations in, 65, 265 n88; and Pennsylvania, 67–71, 73, 95–97, 101, 106–7, 254–55; and New Jersey, 116, 120–21, 126–27, 151, 158; enlargement, 170, 178, 192, 205; and canal cycles, 176–78; financing, 186–87; and branches, mileage and investment (table), 211; and transport costs, 227–28; competition with railroad, 230, 287 n32; counties on, 236, 289 n53; ton-mileage on, 242
Erie Extension Canal, Pennsylvania, 202, 244, 280 n32; investment (table), 246

Erie Railroad, 112
Erie route, 24–39, 41, 47, 55, 264 n84
Erosion, 3
Essex County, N.J., 130, 141
Exploitative canals, *see* Canals
External price levels, relationship between internal and, 180

Factor payments, 232–33
Facts and Arguments (railroad pamphlet), 78, 82–84, 88–90, 95, 267 n26, 268 n39
Fallacy of direct imputation, 226, 230
Farmington Canal, 178
Federal aid, *see* Federal government
Federal government: role in canal-building, 5–6, 19, 24, 113, 193; and Erie Canal, 24, 34–39, 42–47, 56, 60
Federal transportation plan, 33
Federalist (Trenton, N.J.), 150, 275 n27
Federalist Party, 39, 40, 52, 53
Finance: and canals, 146, 186–87, 193; and canal cycles, 179–81
Foreign capital, 35–36, 230; and canal cycles, 183, 186–90, 191–92, 201, 203, 279 n15, 279 n19, 280 n29
Foreign loans, 49, 51, 59, 60–61, 180–81, 263 n77, 283 n64
Foreign markets, 218, 251
Foreign money markets, 180–81
Forman, Joshua, and Erie Canal project, 30–31, 34, 38–39, 43, 259 n27–28
Fort Stanwix, *see* Rome, N.Y.
France, 17–18, 33, 65; *see also* Languedoc Canal
Franklin, Benjamin, 25
Fredonian (New Brunswick, N.J.), 120, 124
Free Banking Act (New York), 191
Fulton, Robert, 1–4, 10, 43, 135, 256 n1–2, 256 n5
Fur trade, 17–18

Gallatin, Albert, 25, 36, 55, 221, 223; *Report*, 5–8, 34–35, 37–38, 43–44, 117, 217–18, 257 n15
Gallman, Robert E., 206, 229–30, 287 n31
Galveston and Brazos Canal, mileage and investment (table), 213

Index

Gazette (Pittsburgh), 94
Geddes, James, and Erie Canal project, 27–28, 31–32, 40, 59, 259 n24, 259 n26, 262 n72
Genesee Messenger (Canandaigua, N.Y.), 28, 30
Genesee River, 18, 31, 41
Genesee Valley Canal, 37; and third canal cycle, 192, 203, 205; mileage and investment *(tables)*, 211, 246
Geneva, N.Y., 24
Gentlemen's Magazine (London), 77
Georgia, 35, canal mileage, investment, and income *(tables)*, 212, 233
German Flats, 22
Germany, railroads in, 85
Gold specie standard, 181
Goodrich, Carter, 1, 24–29, 215, 249, 257 n16, 260 n34, 265 n88
Government intervention, attitudes toward, 161–66; *see also* Federal government; State government
Great Britain: canals in, 1, 2, 4, 9, 63; railroads in, 154; financing U.S. canals, 179–81, 191–92, 280 n29; and third canal cycle, 202; *see also* England
Great Lakes, 15, 41; and Ohio-Mississippi river system, 8, 17, 18, 19
Gulf of Mexico, 17, 37

Harrisburg, Pa., 148, 149
Hartz, Louis, 98–100, 215, 270 n85
Harvey's Canal, mileage and investment *(table)*, 213
Hawley, Jesse, 28–30, 34–38, 40–42, 259 n26
Hercules papers, 28–30, 32, 35–36, 37
Holgate, Jacob, 72
Hocking Canal and branch, 237, 244, 281 n33; mileage and investment *(tables)*, 212, 246
Holland Land Company, 31, 50, 56, 261 n54, 262 n66, 263 n26
Hosack, David, 259 n24, 259 n28, 261 n60, 263 n79
Hudson River, and New York waterways, 4–6, 15–17 *passim;* 20–23 *passim,* 29, 30, 67, 70
Hydraulic cement, 65

Illinois, 6, 45; canal financing in, 194, 197, 278 n12; panic of *1839,* 198–200; canal mileage, investment, and income *(tables)*, 213, 233
Illinois and Michigan Canal, 6, 8; and canal cycles, 177–78, 204, 283 n66; mileage and investment *(table)*, 213; and trade, 230–31
Illinois River, 46
Illinois Waterway, 9
Inclined plane: in canal building, 30, 41, 49, 52, 53, 58; in railroads, 110–11
Income, per capita, 222, 223, 285 n13
Indiana, 6, 45; and canal cycles, 171, 177; canal financing in, 193–94, 281 n36; panic of *1839,* 197, 198–201, 282 n50; canal mileage, investment, and income *(tables)*, 213, 233, 246; *see also* Wabash and Erie Canal
Indians, 17, 33
Inflation, and canal cycles, 182, 190, 207; and economic development, 224
Ingersoll, Charles J., 82
Inland navigation project, *see* Erie Canal
Insurance companies, 5, 179, 195
Intercoastal Waterway, 9
Interest rates, 197, 205
Internal improvements, 1, 64, 80; *see also* Canals; Railroads; *specific canals, e.g.,* Erie Canal; *specific states*
Internal price levels, relationship between external and, 180
Interregional canals, and canal cycles, 176–77
Interregional trade, 228–32
Intraregional canal projects, and canal cycle, 176–79, 277 n9
Intraregional carriers, 225, 228
Investment: in canals, 170–71, 277 n10; in Canada, 179; by state and municipal governments, 179–81, 213–14, 214 *(table)*; concentration in New York, 203–4; primary impact, 205–6; secondary impact, 205–6; estimates of *(table)*, 208–10; relation to transport costs and consumption, 219; and state income *(table)*, 233; criteria for, 239–46; in canal failures, 246

298 Index

Irish, 53, 54, 63
Iron, in New Jersey, 59, 126–27, 135, 146, 160; and railroads, 86
Iron and steel industry, 234
Irondequoit Valley, 32
Isard, Walter, 112, 170, 172, 273 n123, 276 n2

Jackson, Andrew, 188, 196
James River and Kanawha Canal, 5, 6, 8, 204; and canal cycles, 177–78, 285 n17; mileage and investment (tables), 212, 214; and interregional trade, 228
Jefferson, Thomas: on federal subsidy, 34, 36–39 passim, 43; on comparative capability, 65–66, 114
Jersey City, N.J., 130, 136
Jones, Chester Lloyd, 242
Junction Canal, mileage and investment (table), 211
Juniata Division Canal, mileage and investment (table), 212
Juniata Valley, 7

Kent, James, 63, 65
Kentucky, canal mileage and investment (table), 213
Kirkland, Edward Chase, 215, 260 n43
Kirkpatrick, William, 43
Kneass, Samuel, 71

Labor: and Erie Canal project, 3, 4, 31; and Pennsylvania Mainline, 133; and canal cycles, 174, 206, 224
Lake Champlain, 5, 19
Lake Erie: and New York, 4, 7, 18, 31, 32; and Pennsylvania, 73
Lake Michigan, 37, 46
Lake Oneida, 24
Lake Ontario, and transportation, 5, 16–18, 20, 21, 23, 26, 31, 32, 41
Lake Superior, 9, 37
Lancaster County, Pa., 70
Land: speculation, 23, 190, 223–24, 230, 285 n14; grants, 45, 134, 177, 193, 281 n36; public, 61, 190; certificates, 199; values, 220, 234–35, 284 n10
Languedoc, Canal of, 1, 3, 29–30, 65
Lateral canal lines, 175–76, 178, 225
Latrobe, Benjamin, 25

Lebanon County, Pa., 70
Lehigh Coal and Navigation Canal, 6, 242, 277 n9; mileage and investment (table), 211
Lehigh County, Pa., 70
Lehigh Valley, 116, 127, 135, 136, 146
Leonardo da Vinci, 2
Lieber, Francis, 239
Little Falls, N.Y., 21, 22
Liverpool and Birmingham Railroad, 76
Liverpool and Manchester Railroad, 76, 79, 80, 95
Livingood, James Weston, 242
Livingston, Robert R., 43
Loans, 179–81, 186–89, 277 n11, 279 n15
Lockhart's Canal, mileage and investment (table), 212
Locks, and Erie Canal, 2, 25, 30, 49, 59; and Pennsylvania Mainline, 73
Locomotive, 69, 75, 85, 89, 112
London money market, 179–80, 198, 207, 278 n14
Long Island, 57
Lorick's Canal, mileage and investment (table), 212
Loring, N. H., 101
Louis XIV, 1
Louisiana, canal mileage and investment (table), 213
Louisville and Portland Canal, 8–9; mileage and investment (table), 213
Lowe, Enoch L., governor of Maryland, 204

M'Culloch, George P., campaign for New Jersey canal, 126–31
McLane, Louis, 146
McWhorter, John, 30, 34
Madison, James, 45, 51, 60, 261 n49
Maine, canal mileage and investment (table), 211
Mainline, see Pennsylvania Mainline
"Mammoth Plan," 139–42, 275 n27
Manchester and Liverpool Railroad, 76, 79, 80, 95
Marcy, W. L., 192
Maryland: improvement policies of, 6, 98; Chesapeake Bay transport, 67; canal financing in, 194; panic of 1839, 198; canal mileage, invest-

Index

ment, and income (tables), 212, 233
Massachusetts, 47, 98, 177, 178; canal mileage and investment (table), 211
Matthews, R. C. O., 191
Merrimack Valley, 7
Meyer, Balthasar Henry, 215
Miami and Erie Canal and branches, 178, 193, 230–31, 237, 281 n33; mileage and investment (table), 212
Michigan, 6, 45, 47; canal mileage, investment, and income (tables), 213, 233
Middlesex Canal, 6, 7, 59, 177
Milan Canal, government aid to (table), 214
Miller, Nathan, 23, 186, 188, 258 n12–13, 258 n15
Miner's Journal, 84
Mississippi River: and Great Lakes, 8, 17, 37; route to New Orleans, 68, 75, 107, 231
Mixed enterprise, 6, 10; advocated in New Jersey canal project, 139–42; Virginia plan of, 153; and canal cycles, 179, 186, 215 (table); public aid in (table), 214
Model: of canal financing, 181–83; of United States economy, 221–25, 235
Mohawk River, and New York water system, 5, 16–23 *passim*, 29, 30, 67, 70
Mohawk valley, 33
Monetary system, U.S., 188
Money markets, 180, 188, 195–97
Money supply, 190–91, 283 n58
Monongahela Navigation Canal, government aid to (table), 214
Monopoly, 5, 36, 155, 156
Montreal, 17, 18, 57, 264 n84
Morris, Gouverneur, and Erie Canal, 26–28, 34, 37, 40, 45, 52, 58–59, 259 n22–23, 263 n72
Morris and Essex Railroad, 147
Morris Canal: and decision-making, 10–11, 116–46; operations summarized, 146–47, 288 n46; economic impact of, 159–60; mileage and investment (table), 211

Morris Canal and Banking Company, 195, 197, 281 n40
Morris County, N.J., 130, 141
Mud Creek, 23, 31
Multiplier-accelerator effects, 223
Muncy Canal, mileage and investment (table), 212
Municipal investment, 6
Museum (Rahway, N.J.), 128
Muskingum Canal, 237, 281 n33: mileage and investment (table), 213

National Advocate (New York), 63, 122
National bank, 46
National commodity output, on the Erie Canal, 228–32
National Gazette (Philadelphia), 76
National Road, 68, 272 n102
Newark, N.J., 116, 130, 136
Newark Turnpike Company, 115, 274 n2
New Brunswick, N.J., 148
New England, 33, 35
New Hampshire, canal mileage and investment (table), 211
New Haven and Northampton Canal, mileage and investment (tables), 211, 214
New Jersey: private subsidy in, 6, 115, 155, 160–66, 274 n2–3; canals, 10, 11, 115–66 (*see also* Delaware and Raritan Canal; Morris Canal); legislature, 116–19, 121–26, 129, 138–39, 143–44, 150–52, 154; products, 126–27, 135, 146, 159; state subsidy in, 132–33, 160; sectional rivalries in, 134–38, 153, 165–66; banking companies, 162–63; canal mileage, investment, and income (tables), 211, 233
New Jersey Journal, 144
New Orleans, 6, 37, 68, 231
New York (city): access to Midwest, 16, 17; and Erie Canal project, 34, 37, 52–54, 62–63; economic advantages of, 64–65, 112, 273 n123; and turnpike era, 67; and Pennsylvania railroads, 92; and New Jersey produce, 127; money market, 187; prices in, 227

New York (state), 6, 9, 10, 22; early history, 16–19; westward expansion in, 33–34, 69–70; and Pennsylvania, 38–39, 112–14; rivals, 38–39, 68–69, 265 n7; and federal aid, 44–47, 113; and Erie Canal (*q.v.*), 47–52, 57–58, 112; natural advantages, 67–69; social structure in, 70, 264 n82, 266 n9; railroads, 112, 273 n123; and New Jersey canals, 116, 122–23, 145–46; investment in canals, 170–71, 203–4; subsidiary canals in, 175–79 *passim;* route to Ohio River, 176; panic of *1839,* 198–99; canal financing in, 192, 195; canal mileage, investment, and income (*tables*), 211, 233, 246
New York Canal Commission, 38, 196
New York Central Railroad, 112
Niagara Falls, 5, 27, 30
North, Douglass C., 191, 279 n15
North, William, 40
North Branch Extension Canal, 204, 244, 280 n32; mileage and investment (*tables*), 212, 246
North Carolina, canal mileage and investment (*table*), 212
Northern Inland Lock Navigation Company, 21

Ogeechee Canal, mileage and investment (*table*), 212
Ohio, 6, 8, 47; canals in, 96, 171, 177–79, 193, 281 n33; panic of *1839,* 198–201; canal mileage, investment, and income (*tables*), 212–13, 233, 246; employment in, 236–38; *see also* Ohio and Erie Canal
Ohio and Erie Canal: influence, 174, 176, 178, 224, 230–31, 235–38; counties, 237; mileage and investment (*table*), 213
Ohio and Pennsylvania Canal, mileage and investment (*tables*), 213, 214
Ohio Life Insurance and Trust Company, 195
Ohio River, 8–9, 19, 68, 176, 231
Ohio Valley, 57, 176–79
Oneida Lake, 16, 17, 31, 39, 55
Oneida Lake Canal, mileage and investment (*table*), 211
Oneida River, 16

Onondaga County, N.Y., 30, 33–34
Onondaga River, 21
Ontario route, in Erie Canal proposals, 16–25, 40–41, 47, 54–55, 58, 64
Orleans Bank Canal, mileage and investment (*table*), 213
Oswego, N.Y., 20, 32, 55
Oswego Canal, 178, 242; mileage and investment (*table*), 211
Oswego Falls, 21
Oswego River, 16–18

Palladium of Liberty (Morristown, N.J.), 128, 137
Palmyra, N.Y., 31
Panic of *1839,* 195–96, 197–201
Passaic Valley, 137
Paterson, N.J., 127, 130, 136
Pawtucket Canal, mileage and investment (*table*), 211
Peabody, George, 195, 197
Penn, William, 117
Pennine range, 4
Pennsylvania, 6, 8, 10; roads in, 19, 57, 67–68; and New York, 38–39, 69–71, 101, 112–14; drive for transportation to west, 67–69; canal-railroad debate, 69–102, 266 n8; legislature, 82, 92–94; improvement policies of, 82, 98, 149, 268 n36; and railroads, 105–6, 112–14, 273 n123; democratic social structure in, 113; canals, 171, 175, 177–79, 193, 280 n21; and panic of *1839,* 198–201; credit failure, 202–3, 251; canal mileage, investment, and income (*tables*), 211, 233, 246; *see also* Pennsylvania Mainline
Pennsylvania and Ohio Canal, government aid to (*table*) 214
Pennsylvania Mainline: and decision-making, 66, 67–114, 266 n15, 268 n44; contrasted with Erie Canal, 101, 104, 106–7, 112–14, 254–55; cost of, 104, 107, 113, 271 n97; summary of operations, 105–14, 225, 228, 231–32, 271 n97, 272 n101–3, 285 n17; and canal cycles, 176–78; mileage and investment (*table*), 212; ton-mileage on, 242
Pennsylvania Railroad Company, 8, 105–6, 109

Index

Pennsylvania Society; founded, 70–71, 266 n10; in canal-railroad debate, 77–81 *passim*, 86–90 *passim*, 220, 270 n64
Peoria, Ill., 232
Perry County, Pa., 70
Philadelphia, 71, 96; and turnpikes, 57, 68; and railroads, 112; and New Jersey canals, 147–48
Pintard, John, 117
Pittsburgh, 57, 67–68
Pittsburgh Pike, 57
Planning, *see* Economic development; Federal government; State government; *specific canals, e.g.,* Erie Canal
Platt, Jonas, 39, 53, 64
Poland, railroads, 85
Politics: and Erie Canal, 5, 15–66 *passim;* and Pennsylvania Mainline, 67–114 *passim;* and New Jersey canals, 115–66 *passim;* and lateral canal lines, 178–79; and canal cycles, 181–82, 204
Pope, John, 44
Population, *see* Settlement; West
Portage, 91; railroad, 8, 91, 93, 103–5, 111
Porter, James M., 101
Porter, Peter Buell, 40, 43–44
Portsmouth, Ohio, 174
Potomac Canal, to the Ohio, 71–72
Potomac River, 5, 8, 19
Press, in canal-railroad debate, 76–98 *passim;* and New Jersey canals, 120–53 *passim*
Pressure groups; *see* Politics; Press; Sectional rivalry
Price index, 232
Prime, Ward and King, 187, 195
Prisms, 170, 277 n5
Private enterprise, 4–6; and canal building, 9, 21–25, 35, 42, 60, 115–66 *passim*, 159–60, 177, 179–80, 186, 253–54, 258 n8, 258 n11; and railroads, 105–6
Public financing, in first canal cycle, 186
Public investment, in canals (*table*), 215
Public works programs, *see* Canals; Federal government; State government
Puddling, 3

"Radical" Democrats, 204
Railroads: and canals, 2, 9–11, 15, 53, 64, 69–102 *passim*, 152–54, 183, 208, 228–30, 244–46, 270 n64, 287 n32; in England, 75–77, 79–81, 84–85, 87–88, 94–95; technological developments, 75–76, 80, 83, 85, 87, 89–90, 109–112; for portage, 91, 103–5; private construction, 105–6; *see also* Pennsylvania Mainline; *specific names, e.g.,* Pennsylvania Railroad
Rate-cutting, 23
Raymond, Daniel, 219–20, 284 n7–8
Recessions: of *1825*, 188; of *1829*, 188–89; of *1833–34*, 188–89
Regional price differentials, 222–25, 227
Regional specialization, 225
Renwick, James, 130, 135–37
Republican Party, 39–40
Reservoirs, 3
Revolutionary War, 17, 18, 19, 26
Rhode Island, 178; canal mileage and investment (*table*), 211
Richmond, Va., 177
Riquet, M., 3
Rivalry, *see* Sectional rivalry
River improvements, 3, 21–26, 28, 40; *see also* Ontario route
Robinson, Moncure, 103
Rochester, N.Y., 232, 262 n70
Rome (Fort Stanwix), N.Y., 18, 19–22 *passim*, 49
Rubin, Julius, 15, 67, 265 n7, 266 n10, 268 n39, 269 n56
Rutherford, John, 134

Sadove, Abraham H., 231, 272 n101–2, 288 n40–41
St. Lawrence, 17, 30, 32
St. Marie straits, 37
St. Mary's Falls (Soo) Canal, 9; mileage and investment (*table*), 213
Saluda Canal, mileage and investment (*table*), 212
Sanderlin, Walter S., 215
Sandy and Beaver Canal, mileage and investment (*table*), 213
Sandy Hook, 117
Santee Canal, 6, 7, 177

Santee River, 5
Savannah River, 5
Schenectady, N.Y., 17, 22, 29
Schlatter, Charles L., 105
Schuyler, Philip, 20, 21, 25–26, 258 n9
Schuylkill Navigation Canal, 242; mileage and investment (*tables*), 212, 214
Schuylkill River, 67, 70
Scotland, 77
Sectional rivalry, 176–79, 181, 251–52
Segal, Harvey H., 169, 215, 217, 271 n97, 276 n3, 277 n6
Seneca Falls, 21
Seneca Lake, 17, 20, 21, 41, 49
Seneca River, 16, 18, 21, 31, 39
Sentinel (Utica, N.Y.), 120
Settlement, 11, 207, 217, 230; and population, 34, 227, 285 n14, 285 n16, 289 n53; *see also* West
Sewall's Falls Canal, mileage and investment (*table*), 211
Seward, William H., 192, 199
Shipping routes, 230–32
Shulze, John A., 92–94, 219, 223
Smith, Adam, 217, 220
Smith, Walter Buckingham, 179–80, 278 n145
Society for the Establishment of Useful Manufactures, 137
Soo (St. Mary's Falls) Canal, 9; mileage and investment (*table*), 213
South Carolina, 177; canal mileage and investment (*table*), 212
Specie, in banks, 180
Speculation: land, 23, 190, 223–24; and Morris Canal, 145–46; in banks, 163; and canal cycles, 183, 207, 223
Stagger period, and canal cycles, 175–76
State government, 5–6, 19; and Erie Canal, 13–66 *passim*, 158; and Pennsylvania Mainline, 67–114 *passim;* and New Jersey canals, 120–26, 132–33, 137–38, 148–53, 155; and developmental and exploitative canals, 158; financing of internal improvements, 161–62; borrowing for internal improvements, 179–81, 277 n11; and canal cycles, 182, 192–201; indebtedness of, 201–3; percent of canal investment, 213–14

Steamboat, 68, 75
Stephenson, George, 75, 76
Stevens, John, 74–75, 80, 267 n29
Stevens, John Cox, 154
Stevens, Robert, 154
Stevenson, Robert, 77
Stock subscriptions, to finance canals, 179
Stockton, Robert F., 154
Stockton and Darlington Railroad, 76, 77
Strickland, William, 71, 78–79; *Report*, 81, 84, 86–87, 90, 94, 268 n39, 270 n64
Subsistence farming, 222
Surpluses, regional, 217
Susquehanna and Tidewater Canal, mileage and investment (*tables*), 212, 214
Susquehanna Division Canal, 220; mileage and investment (*table*), 212
Susquehanna River, 5, 67, 70, 102
Susquehanna Valley, 7
Sussex County, N.J., 130, 141
Swift, Joseph G., 130
Switzerland, 45
Syracuse, N.Y., 30, 232

Tammany organization, 52–53, 54, 63, 264 n82
Taxes: and Erie Canal project, 61; and New Jersey canals, 130, 147; and canal financing, 180, 183, 203, 207; aversion to, 192, 195, 252; and canal cycles, 205
Technology, and canal cycles, 174–76; *see also* Canals; Railroads
Tennessee, 47
Texas, canal mileage and investment (*table*), 213
Tidewater navigation, 35
Time-series curves, 171, 173
Times (New Brunswick, N.J.), 120, 123
Tocqueville, Alexis de, 169
Toledo, Ohio, 232
Tolls, 240; and Erie Canal, 5, 21–25 *passim*, 36, 48, 60, 61, 187; and Pennsylvania Mainline, 107; and

New Jersey canals, 140, 147, 150, 155
Tompkins, Daniel, 52, 58, 63
Tonnewanta Creek, 31
Trade, 1, 176–79, 228–32, 288 n40
Transport costs: relation to consumption and investment, 219, 221–23, 225; and canals, 227–32, 238, 249, 286 n25; direct benefits of, 241–42
Trenton, N.J., 148, 154
Troy, N. Y., 22
True American (Trenton, N.J.), 120, 122
Tunnels, 72–74 *passim*, 103
Trcziyulney, Charles, 73, 74
Turnpikes, 23, 35, 57; in Pennsylvania, 67–68, 71, 108; 272 n101; in Virginia, 177; trade on, 230–31

Underutilization of canals, 244
Union Canal, 70, 75, 100, 102; mileage and investment *(tables)* 212, 214; ton-mileage on, 242
Union County, Pa., 70
United States Gazette, 77–79 *passim*, 83, 88, 95, 266 n15
United States government, *see* Federal government
United States Military Academy, 4
Urban growth, and canals, 224, 225, 232
Utica, N.Y., 232

Van Buren, Martin, 59, 61, 62, 102
Van Rensselaer, Stephen, 40, 60, 263 n75
Vermeule, Cornelius C., 274 n3
Vethake, Henry, 220, 234, 284 n9
Virginia, 6, 67; canals in, 194, 204; canal mileage, investment, and income *(tables)*, 212, 233

Wabash and Erie Canal: importance of, 6, 8, 230–31; and canal cycles, 177–78, 193–94, 204, 281 n36, 283 n66; mileage and investment *(tables)*, 213, 246; competition with railroad, 245
Wagon transportation, 1, 227–28, 231
Wales, 77

Walhonding and Mohican Canal, 244, 281 n33; mileage and investment *(tables)*, 213, 246
Wando Catawaba Canal, government aid to *(tables)*, 214
War of *1812*, 6, 51, 64, 118, 264 n84
Warren Canal 281 n33
Washington, George, 19, 43
Washington, George C., 197
Wateree Canal, mileage and investment *(tables)*, 212, 214
Water rights, 147–48
Watson, Elkanah, 19–21, 26, 28, 36–37, 258 n5–9
Weaver, Charles Clinton, 215
Weldon Canal, mileage and investment *(table)*, 212
West, 33–34, 221–22, 223, 227–32, 272 n102
West Branch Canal, 244, 280 n32; mileage and investment *(tables)*, 212, 246
Western Division Canal (Pennsylvania Mainline), mileage and investment *(table)*, 212
Western Inland Lock Navigation Company, and Erie Canal, 21–25, 36, 39–40, 43, 50
Weston, William, 22–23, 26, 49, 52, 58, 261 n60
Wheeler, Thomas, 34
Wheeling, W. Va., 57, 67, 68
Whigs, 192
Whitewater Canal, 202, 244; mileage and investment *(tables)*, 213, 246
Whitford, Noble E., 52, 61–62, 226–27, 261 n60, 263–64 n79, 285 n19, 286 n20–21
Williams, Micajah T., 187
Williamson, Hugh, 38, 260 n37
Wilson, John, 103
Winyaw Canal, government aid to *(table)*, 214
Wisconisco Canal, 244; mileage and investment *(tables)*, 212, 246
Wolf Rift, 22
Wood Creek 16, 19, 21, 22
Wordsworth, William, 202, 283 n61
Wright, Benjamin, 38, 59, 130, 259 n28, 263 n72
Wright, Silas, 239